Control and Automation

Control and Automation

Editors

Cristina I. Muresan
Eva H. Dulf

MDPI • Basel • Beijing • Wuhan • Barcelona • Belgrade • Manchester • Tokyo • Cluj • Tianjin

Editors

Cristina I. Muresan
Faculty of Automation and
Computer Science, Department
of Automation, Technical
University of Cluj-Napoca
Romania

Eva H. Dulf
Faculty of Automation and
Computer Science, Department
of Automation, Technical
University of Cluj-Napoca
Romania

Editorial Office
MDPI
St. Alban-Anlage 66
4052 Basel, Switzerland

This is a reprint of articles from the Special Issue published online in the open access journal *Applied Sciences* (ISSN 2076-3417) (available at: https://www.mdpi.com/journal/applsci/special_issues/Control_Automation).

For citation purposes, cite each article independently as indicated on the article page online and as indicated below:

LastName, A.A.; LastName, B.B.; LastName, C.C. Article Title. *Journal Name* **Year**, *Volume Number*, Page Range.

ISBN 978-3-0365-2446-7 (Hbk)
ISBN 978-3-0365-2447-4 (PDF)

Contents

About the Editors

Cristina I. Muresan received a degree in control systems and a Ph.D. degree from the Technical University of Cluj- Napoca, Cluj-Napoca, Romania, in 2007 and 2011, respectively, where she is currently an Associate Professor in the Automation Department. She has published more than 150 papers and books. She is currently involved in several research grants, all dealing with multivariable, biomedical engineering and fractional order control. Her research interests include modern control strategies, such as predictive algorithms, robust non-linear control, fractional order control, time delay compensation methods, and multivariable systems.

Eva H. Dulf received her Ph.D. degree in Advanced Process Control, with applications in nuclear technologies, from the Technical University of Cluj-Napoca, Romania, in 2006, where she is currently a professor and leader of the Advanced Process Control Methods Group. She obtained her habilitation degree in 2015. Her research interests include advanced control strategies. More specifically, her research focuses on developing advanced control strategies for complex chemical and biochemical processes. She has published over 100 papers in journals and conferences and several books or book chapters. She has won 34 awards at international exhibitions of inventions. She has been an IEEE member since 2009.

Preface to "Control and Automation"

Control and automation systems are at the heart of our everyday lives. This book is a collection of novel ideas and findings in these fields, published as part of the Special Issue on Control and Automation. The core focus of this issue was original ideas and potential contributions for both theory and practice. It received a total number of 21 submissions, out of which 7 were accepted. These published manuscripts tackle some novel approaches in control, including fractional order control systems, with applications in robotics, biomedical engineering, electrical engineering, vibratory systems, and wastewater treatment plants.

One of these papers presents a synchronous co-simulation of a six degree of freedom (6DOF) ball and plate platform, and its 3D computer model intended to simulate the actual dynamics of a rendezvous between a cargo vehicle, such as the Falcon 9 from SpaceX and the International Space Station (ISS). A supervisory action is required for initiating the docking mechanism. The novelty of the manuscript consists of an adaptive fractional order control solution. The tuning method is easy, and the control strategy is implemented and validated based on a laboratory benchmark. The results demonstrate that such an approach is suitable to meet the performance specifications, despite the large variability in the system dynamics. Vibratory systems and an efficient control for these processes are addressed in two of the published papers. One of these tackles the idea of using a fractional order differential equation to model such a process, based on its viscoelastic nature. Then, a fractional order linear quadratic regulator (LQR) is designed. To solve the Riccati equation of the LQR method, an iteration-based approach is proposed. Next, to estimate the states originating from the fractional-order derivative term, a fractional-order state observer is constructed. To demonstrate the efficiency of the proposed control algorithm, numerical simulations are presented, showing that vibrations are indeed suppressed. The second paper, which deals with vibratory systems, uses a smart beam to validate an experimental tuning approach for fractional order proportional–integral–derivative (PID) controllers. The tuning method is based on shaping the Bode magnitude plot to decrease the resonant peak. There is no need for a process model; only several experimental tests are required to obtain the necessary process information at different frequencies. Experimental results are provided to demonstrate the efficiency of the proposed approach. Both of the abovementioned algorithms use concepts from fractional calculus. To implement such fractional order controllers, digital realizations are often required, which use sample and hold (S&H) circuits to perform the conversion from analog to digital and vice versa. An excellent review of these S&H systems, as well as a new fractional order design of such systems, is presented in this issue. The proposed approach models these systems both in the time and Laplace domains and is a generalization of the classical devices, enabling a better understanding of the possibilities and limitations of S&H systems.

The field of biomedical engineering is currently receiving tremendous attention from the research community, especially regarding personalized and optimized medicine. Such a topic is addressed regarding carcinoma models and adequate chemotherapeutic treatment. The core problem in the delivery of a carcinoma model suitable for developing an optimized therapy consists of a limited number of measurable physiological signals—state variables—and the knowledge of model parameters. The proposed solution refers to the use of observers. A moving horizon estimation (MHE)-based observer is developed that uses a third-order tumor growth model. The performance of the observer is compared to that of an optimized extended Kalman filter (EKF). Results show that the proposed MHE is designed to be suitable for closed-loop applications and yields simultaneous state and parameter estimation.

Novel ideas in the field of optimal battery charging are also presented in this Special Issue. A battery model, designed for automotive battery management systems applications, is used to design charging controllers in an optimal charging strategy. The standard model is altered to produce a non-linear electro-thermal battery aging model, which is linearized at several operating points. An optimal charging strategy is then designed for the simultaneous minimization of charging time and maximization of battery lifetime. An optimization problem that minimizes aging is solved to determine an appropriate charging trajectory. The CRONE methodology is then used to take into account the resulting linear model family. Several simulations are employed to show the efficiency of the resulting charge controller.

The Special Issue contains a seventh paper dealing with a weather-based prediction strategy inside the proactive historian with application in wastewater treatment plants. The authors propose a software reference architecture for the proactive historian, alongside a data dependency identification strategy and several applications for energy efficiency improvements in the water industry. However, complex research is required in order to offer a complete solution for real industrial processes. Part of this reference architecture that predicts the future evolution of the monitored system is developed in one of the papers of this Special Issue. The design is based on weather dependency and forecasting. In this way, the approach paves the way towards achieving a fully functional, real-world, tested, and validated proactive historian application, with potential to bring significant direct benefits to the water industry.

This Special Issue has gathered a selection of novel research results regarding control systems in several distinct research areas. We hope that these papers will evoke new ideas, concepts, and further developments in the field.

We wish to thank the authors for considering this Special Issue as a suitable way to disseminate their results. Additionally, we thank the reviewers for the extra effort put into reviewing the manuscripts. Finally, we thank the dedicated editorial team of *Applied Sciences*, especially assistant editor Wing Wang, the Special Issue Managing Editor, who provided help enormously with managing the papers.

Cristina I. Muresan, Eva H. Dulf
Editors

applied sciences

MDPI

Editorial

Special Issue: "Control and Automation"

Cristina-Ioana Muresan and Eva H. Dulf *

Department of Automation, Technical University of Cluj-Napoca, Memorandumului Street, No 28, 400114 Cluj-Napoca, Romania; Cristina.Muresan@aut.utcluj.ro
* Correspondence: Eva.Dulf@aut.utcluj.ro

1. Introduction

A wide range of ongoing research in the areas of controller design and information engineering reveals that the pace of technological change in this domain seems to be accelerating. The primary focus is on futuristic segments ranging from robotics and genomics to chemical feedstock and electrical storage.

The main objective of this Special Issue was to provide a forum for researchers and practitioners to exchange their latest theoretical and technological achievements and to identify critical issues and challenges for future investigation on topics regarding advanced automation and control techniques based on information system technologies.

2. Novel Ideas for Control

The Special Issue focused on original ideas and potential contributions for theory and practice, at the same time. It received a total number of 21 submissions, among which 7 were accepted. These published manuscripts tackle some novel approaches in control, including fractional order control systems, with applications in robotics, biomedical engineering, electrical engineering, vibratory systems, waste water treatment plants.

One of these papers [1] presents a synchronous co-simulation of a 6DOF (six degrees of freedom) ball and plate platform and its 3D computer model intended to simulate the actual dynamics of a rendezvous between a cargo vehicle such as the Falcon 9 from SpaceX and the ISS (International Space Station). A supervisory action is required for initiating the docking mechanism. The novelty of the manuscript consists of an adaptive fractional order control solution. The tuning method is easy and the control strategy is implemented and validated on a laboratory benchmark. The results demonstrate that such an approach is suitable to meet the performance specifications despite the large variability in the system dynamics.

Vibratory systems and an efficient control for these processes are addressed in two of the published papers. One of these [2] tackles the idea of using a fractional order differential equation to model such a process, based on its viscoelastic nature. Then, a fractional order Linear Quadratic Regulator (LQR) is designed. To solve the Riccati equation of the LQR method, an iteration-based approach is proposed. Next, to estimate the states originating from the fractional-order derivative term, a fractional-order state observer is constructed. To demonstrate the efficiency of the proposed control algorithm, numerical simulations are presented showing that vibrations are indeed suppressed. The second paper that deals with vibratory systems uses a smart beam to validate an experimental tuning approach for fractional order proportional–integral–derivative (PID) controllers [3]. The tuning method is based on shaping the Bode magnitude plot, such that the resonant peak is decreased. There is no need for a process model, only several experimental tests are required to obtain the necessary process information at different frequencies. Experimental results are provided to demonstrate the efficiency of the proposed approach.

Both of the algorithms in [2,3] use concepts from fractional calculus. To implement such fractional order controllers, digital realizations are often required which use sample and hold (S&H) circuits to perform the conversion from analog to digital and vice versa.

Citation: Muresan, C.-I.; Dulf, E.H. Special Issue: "Control and Automation". *Appl. Sci.* **2021**, *11*, 5005. https://doi.org/10.3390/app11115005

Received: 24 May 2021
Accepted: 25 May 2021
Published: 28 May 2021

An excellent review of these S&H systems, as well as a new fractional order design of such systems, is presented in [4]. The proposed approach models these systems both in the time and Laplace domains and is a generalization of the classical devices, enabling a better understanding of the possibilities and limitations of S&H systems.

The field of biomedical engineering is currently receiving tremendous attention from the research community, especially regarding personalized and optimized medicine. In [5], such a topic is addressed regarding carcinoma models and adequate chemotherapeutic treatment. The core problem in delivering a carcinoma model suitable for developing an optimized therapy consists of the limited number of measurable physiological signals—state variables—and the knowledge of model parameters. The solution proposed in [5] refers to the use of observers. A moving horizon estimation (MHE)-based observer is developed that uses a third-order tumor growth model. The performance of the observer is compared to that of an optimized extended Kalman filter (EKF). Results show that proposed MHE is designed to be suitable for closed-loop applications and yields simultaneous state and parameter estimation.

Novel ideas in the field of optimal battery charging are presented in [6]. A battery model, designed for automotive battery management systems applications, is used to design charging controllers in an optimal charging strategy. The standard model is altered to produce a nonlinear electro-thermal battery aging model, which is linearized in several operating points. An optimal charging strategy is then designed for simultaneous minimization of charging time and maximization of battery lifetime. An optimization problem that minimizes aging is solved to determine an appropriate charging trajectory. The CRONE methodology is then used to take into account the resulting linear model family. Several simulations are employed to show the efficiency of the resulting charge controller.

The Special Issue contains a seventh paper dealing with a weather-based prediction strategy inside the proactive historian with application in wastewater treatment plants [7]. The authors propose a software reference architecture for such a proactive Historian, along with a data dependency identification strategy and some obtained recipes for energy efficiency improvements in the water industry. However, complex research is required in order to offer a complete solution for real industrial processes. Part of this reference architecture that predicts the future evolution of the monitored system is developed in [7]. The design is based on weather dependency and forecast. In this way, the approach paves the way towards achieving a fully functional, real-world, tested and validated proactive Historian application, with potential to bring significant direct benefits to the water industry.

3. Conclusions

This Special Issue has gathered a selection of novel research results regarding control systems in several distinct research areas. We hope that these papers will raise new ideas, concepts and further developments in the field.

We wish to thank the authors for considering this Special Issue as a proper way to disseminate their results. Additionally, we thank the reviewers for the extra effort put into reviewing the manuscripts. Last, but not least, we thank the dedicated editorial team of *Applied Sciences*, especially to assistant editor Wing Wang, the Special Issue Managing Editor who provided a lot of help with managing the papers.

Author Contributions: Conceptualization, C.-I.M. and E.H.D.; writing—review and editing, C.-I.M. and E.H.D. Both authors have read and agreed to the published version of the manuscript.

Funding: This research was funded by grant of the Ministry of Research, Innovation and Digitization, CNCS/CCCDI–UEFISCDI, project number PN-III-P1-1.1-TE-2019-0745, within PNCDI III.

Institutional Review Board Statement: Not applicable.

Conflicts of Interest: The authors declare no conflict of interest.

References

1. Copot, C.; Muresan, C.I.; Beschi, M.; Ionescu, C.M. A 6DOF virtual environment space docking operation with human supervision. *Appl. Sci.* **2021**, *11*, 3658. [CrossRef]
2. Takeshita, A.; Yamashita, T.; Kawaguchi, N.; Kuroda, M. Fractional-order LQR and state observer for a fractional-order vibratory system. *Appl. Sci.* **2021**, *11*, 3252. [CrossRef]
3. Birs, I.; Folea, S.; Prodan, O.; Dulf, E.; Muresan, C. An experimental tuning approach of fractional order controllers in the frequency domain. *Appl. Sci.* **2020**, *10*, 2379. [CrossRef]
4. Ortigueira, M.D.; Machado, J.T. A review of sample and hold systems and design of a new fractional algorithm. *Appl. Sci.* **2020**, *10*, 7360. [CrossRef]
5. Siket, M.; Eigner, G.; Drexler, D.A.; Rudas, I.; Kovács, L. State and parameter estimation of a mathematical carcinoma model under chemotherapeutic treatment. *Appl. Sci.* **2020**, *10*, 9046. [CrossRef]
6. Mohajer, S.; Sabatier, J.; Lanusse, P.; Cois, O. Electro-thermal and aging lithium-ion cell modelling with application to optimal battery charging. *Appl. Sci.* **2020**, *10*, 4038. [CrossRef]
7. Nicolae, A.; Korodi, A.; Silea, I. Weather-based prediction strategy inside the proactive historian with application in Wastewater Treatment Plants. *Appl. Sci.* **2020**, *10*, 3015. [CrossRef]

applied sciences

MDPI

Article

A 6DOF Virtual Environment Space Docking Operation with Human Supervision

Cosmin Copot [1,*], Cristina I. Muresan [2], Manuel Beschi [3] and Clara M. Ionescu [4,5]

[1] Department of Electromechanics, University of Antwerp, OP3Mech, Groenenborgerlaan 171, 2020 Antwerp, Belgium
[2] Department of Automatic Control, Technical University of Cluj Napoca, Memorandumului 28, 400114 Cluj-Napoca, Romania; cristina.muresan@aut.utcluj.ro
[3] Department of Mechanical and Industrial Engineering, University of Brescia, Via Branze 38, 25123 Brescia, Italy; manuel.beschi@unibs.it
[4] Department of Electromechanical, Systems and Metal Engineering, Ghent University, Tech Lane Science Park 125, 9052 Gent, Belgium; claramihaela.ionescu@ugent.be
[5] Flanders Make, EEDT Core Lab on Decision and Control, Tech Lane Science Park 131, 9052 Gent, Belgium
* Correspondence: cosmin.copot@uantwerpen.be

Abstract: In this work, we present a synchronous co-simulation of a 6DOF (six degree of freedom) ball and plate platform and its 3D computer model. The co-simulation in the virtual environment is intended to mimic the rendezvous between a cargo vehicle such as the Falcon 9 from SpaceX and the ISS (International Space Station). The visual feedback sensing of the position of the 6DOF platform is implemented using a Kinect RGB-D device. The human in the loop acts as supervisory control for initiating the docking mechanism. This paper delivers an adaptive fractional order control solution which is easily tunable, implementable and validated on a laboratory benchmark. The results indicate that fractional order control can tackle large variability in the system dynamics and deliver specified performance at all times.

Keywords: mechatronic system; 6DOF platform; co-simulation platform; robust control; fractional order PID control; linear parameter varying system

Citation: Copot, C.; Muresan, C.I.; Beschi, M.; Ionescu, C.M. A 6DOF Virtual Environment Space Docking Operation with Human Supervision. *Appl. Sci.* **2021**, *11*, 3658. https://doi.org/10.3390/app11083658

Academic Editor: Enrico Vezzetti

Received: 3 December 2020
Accepted: 9 April 2021
Published: 19 April 2021

1. Introduction

There are many possible applications for the six degree of freedom (6DOF) ball and plate platform, ranging from flight simulators, gaming and mimicking space rendezvous for docking and berthing mechanisms. Space engineering applications have increased degrees of LPV (linear parameter varying) dynamics due to their size, relative position, remote location and high instrumentation complexity [1]. The mechanical design of such systems may not be optimal from a control point of view, but rather from a practical point of view (e.g., remote access, testing, validation protocols, etc.). Thus, the control task becomes more challenging within the limitations imposed by the hardware and context [2–4].

Space rendezvous missions are characterized by a high degree of uncertainty in relative inertial load and relative mass distribution depending on the angles of contact between the meeting parts. For example, a leg has to put less effort into accelerating the platform in a vertical translation when the platform is at maximal height than when the platform is at its lowest level. Since the experienced loads do not correspond to tangible masses, the experienced load is referred to as a reflected mass. Therefore, the load on top of an LEMA (linear electro-mechanical actuator) is a non-fixed parameter. Since the dynamic models corresponding to these variable reflected masses change significantly, the variation of the dynamics should be taken into the consideration during the controller tuning process. A specific application is introduced in this paper, as it has the advantage to mimic real-life conditions in a lab scale environment. The challenge of model uncertainty has been addressed previously with a model-based control scheduling scheme and a robust control algorithm [5–7].

This paper presents a co-simulation between a real world model and a virtual world model of a Stewart platform simulator. The purpose is to combine real-world parts (a 6DOF platform) with a virtual model (an equivalent docking 6DOF platform) and mimic a space rendezvous. There is no one-controller-fits-all solution available due to the various orders of magnitude of the difference between dynamic values of the system operation. An adaptive control algorithm ensures the system can abide to aforementioned performance specifications in the presence of high LPV conditions.

The paper is organized as follows. The real world and virtual environment setup, afferent software and hardware facilities and the mimicking co-simulation of the virtual-reality part of the system are described in Section 2. Next, an adaptive controller is proposed in Section 3, allowing human-in-the-loop interaction. The results and discussion are given in Section 4, and Section 5 presents the conclusions and summarizes the main outcome of this work.

2. Real and Virtual Setup Description

The Berthing Docking Mechanism (BDM) is a mechatronic system that is capable of performing successful rendezvous/contact operations via either docking or berthing. It is designed for the low-impact docking of a spacecraft by actively reducing the impact forces on the platform during progressive contact. The structural design of the BDM is a 6DOF parallel manipulator, based on the Stewart–Gough platform [8,9]. Lab-scale 6DOF Stewart systems controlled with robust integer-order controllers have already been presented in [10–13]. However, the main difference to the lab-scale application is that BDM structures have high LPV dynamic properties. Since the leg's dynamics depend on the experienced inertial load, each relative mass corresponds to a model with different dynamics. In other words, each model represents the dynamics of a particular leg in a particular pose when a certain force is exerted on the platform's upper ring. Since every executable trajectory of the platform can be seen as a succession of different states, it is clear that the LEMA dynamics change constantly during spacecraft docking. How they become active during the contact process is briefly described as follows. During docking, the mechanism consists of an active and a passive module, as illustrated in Figure 1, which can be viewed as two sub-systems with an interaction when in contact.

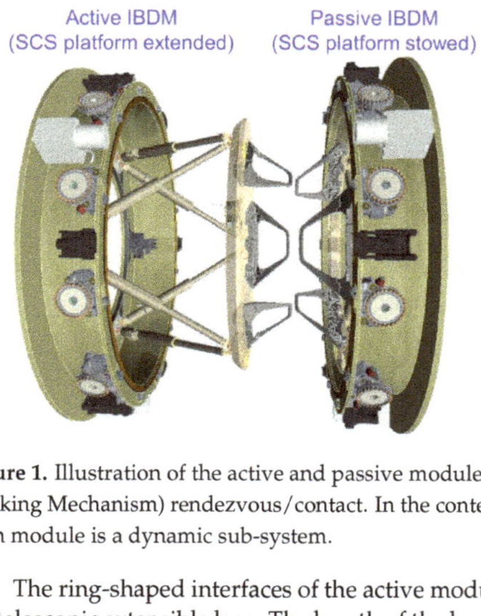

Figure 1. Illustration of the active and passive modules for space IBDM (International Berthing and Docking Mechanism) rendezvous/contact. In the context described in our methodological approach, each module is a dynamic sub-system.

The ring-shaped interfaces of the active module can be moved with 6DOF by means of six telescopic extensible legs. The length of the legs is controlled for each leg individually by a brushless DC motor, based on the force signals measured by the load cells. Considering the platform's 6DOF, the net power the legs have to deliver in order to move the spacecraft in a certain direction might differ, depending on the pose of the platform. When the actuator is oriented in the direction in which the platform has to be accelerated, the legs will experience low-load behavior, which is less demanding in terms of actuator effort. The variations in the load are referred to as the reflected mass. From an LPV perspective, this 6DOF platform has to be controlled with respect to a varying-load system conditions.

The real picture of the platform is depicted in Figure 2a, while the coordinate system of the virtual and real platform is illustrated in Figure 2b. In this figure, the O_{vw} is the origin of the virtual world, the O_{vwp} is the central coordinate of the virtual platform and O_{rw} and O_{rwp} represent the coordinates of the real world and of the real platform, respectively. $H_{rp} = H_{vp} = 50$ cm represents the height of the platform, $D = 10$ cm is the distance between the real ad the virtual platform, and the diameter of the platform plate is 70 cm. The real 6DOF platform from Figure 2a represents the first part of the co-simulation setup system and is considered to be the active module of the BMD setup. It uses the position control of each leg (Figure 3a) to achieve global positioning and determine the angle of the platform.

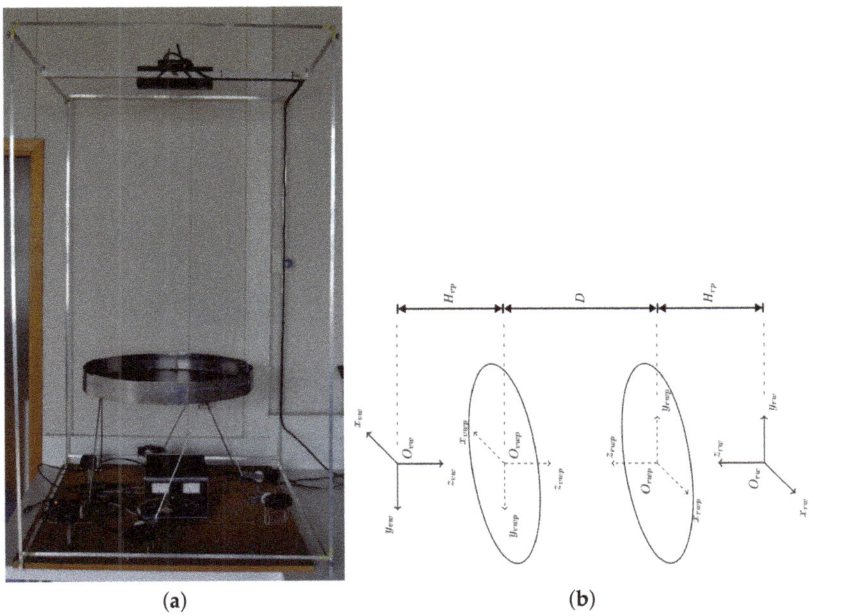

Figure 2. (**a**) Real world six degree of freedom (6DOF) platform for co-simulation, including a frame and Kinect IR-camera; (**b**) Relation between coordinate systems of the real world and the virtual world.

Given that the position of the platform is known, the position of each connection $C(x_c, y_c, z_c)$ is also known (as specified by the controller). The position of the servo $A(x_a, y_a, z_a)$ and its orientation θ are determined by the motion platform construction. The connection of the servo lever and the vertical rod is indicated as $B(x_b, y_b, z_b)$. Let r be the length of the servo arm (AB) while BC is the shaft that connects the servo arm with the platform in the point C and has the length ℓ. The absolute distance between point A and C is denoted by ℓ_1. The inverse kinematic model is described as

$$\varphi = \pm \arccos\left(\frac{M}{\sqrt{K^2 + N^2}}\right) + \arctan\left(\frac{N}{K}\right) + k\pi,\ \forall k \in \mathbb{Z}, \tag{1}$$

where

$$\begin{aligned} M &= \ell^2 - \ell_1^2 - r^2 \\ K &= 2(x_a - x_c) r \cos\theta + x_a + 2(y_a - y_c) r \sin\theta + y_a \\ N &= 2(z_a - z_c) r \end{aligned} \quad . \tag{2}$$

The inverse kinematic model from (1) has four solutions:

$$\begin{aligned} \varphi_1 &= \arccos\left(\frac{M}{\sqrt{K^2+N^2}}\right) + \arctan\left(\frac{N}{K}\right) + \pi \\ \varphi_2 &= -\arccos\left(\frac{M}{\sqrt{K^2+N^2}}\right) + \arctan\left(\frac{N}{K}\right) + \pi \\ \varphi_3 &= \arccos\left(\frac{M}{\sqrt{K^2+N^2}}\right) + \arctan\left(\frac{N}{K}\right) \\ \varphi_4 &= -\arccos\left(\frac{M}{\sqrt{K^2+N^2}}\right) + \arctan\left(\frac{N}{K}\right) \end{aligned} \quad . \tag{3}$$

which are illustrated in Figure 3b. Due to the employed mathematical method, a second point C' is found; thus, solutions $|varphi_3$ and φ_4 (for points B_3 and B_4) are discarded. Solution φ_1 and φ_2 (for points B_1 and B_2) brings the connection C to its desired position. As the servo position is restricted between $\pm 90°$, solution φ_1 (point B_1) is selected as the unique feasible solution. The closed loop control uses classical visual feedback control based on camera vision information, as described in [14].

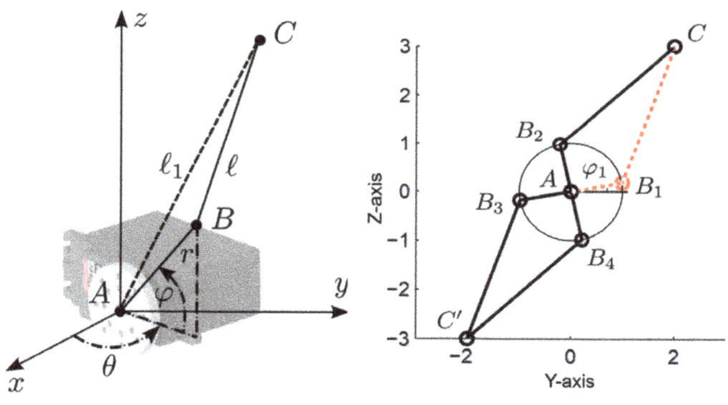

Figure 3. (**a**) Local coordinate frame for a servo control unit; (**b**) mapping of possible solutions for (1).

There are many possibilities for mapping a 3D environment, ranging from very expensive and accurate technologies to low-cost devices which are available for consumers. The Microsoft Kinect is a perfect example of such a low-cost sensor, and this has had a huge impact on recent research in computer vision as well as in robotics. Here, a Kinect v2 sensor is used as it was available in the laboratory. The Kinect v2 sensor holds two cameras (a RGB and an infrared (IR) camera), an IR emitter and a microphone bar, as shown in Figure 4a. The Kinect software maps the depth data to the coordinate system, as shown in Figure 4a, with its origin located at the center of the IR sensor. Note that this is a right-handed coordinate system with the orientation of the y-axis depending on the tilt of the camera and one unit representing one meter. Kinect v2 uses optical time-of-flight (ToF) technology in order to retrieve the depth data with the infrared (IR) camera. The basic principle of ToF is to measure the time difference between an emitted light ray and its collection after reflection from an object (Figure 4b). The RGB-camera within the Kinect sensor has a resolution of 1920 × 1080 pixels with a field of view of 70.6 × 60 degrees, while the IR-camera resolution is 512 × 424 pixels and the field of view is 84.1 × 53.8 degrees. The operative measuring range of the Kinect sensor is between 0.5 and 4.5 m and the frame rate is 30 fps.

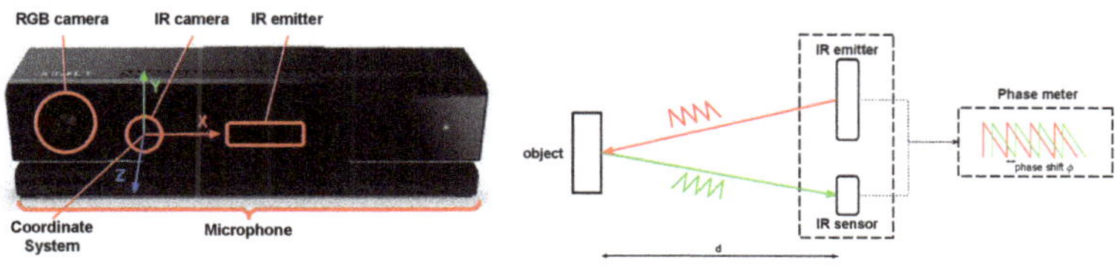

(a) (b)

Figure 4. (**a**) The coordinate system of the Kinect v2 camera and the location of the sensors [15]; (**b**) time-of-flight (ToF) principle used in the Kinect sensor. The sent and received signal are compared to each other in order to determine the phase change and calculate the distance.

The intrinsic parameters of the depth camera of the Kinect, the focal length and the principal point, can be acquired using the Kinect For Windows SDK while the intrinsic parameters of the RGB camera can be acquired with the Camera Calibration Toolbox for MATLAB.

Next, the virtual model was constructed with a Siemens NX12.0 based on the existing 6DOF model of the platform [10–13]. Combining the NX12.0 Motion environment and the inverse kinematics calculations programmed in VisualBasic.Net, a simulation of the virtual world model was executed. Finally, a connection between the LabVIEW program and the NXOpen program was created using registry keys in the Windows Registry, as this allowed for the continuous monitoring of the keys without hindering access to the key from other processes.

In total, 13 VIs were generated to provide the necessary library for communication and the real-time operation of the real world platform and the virtual platform for mimicking a rendezvous. The LabVIEW platform provides the following features for the real-world 6DOF platform:

- Communication with the 6DOF platform;
- Kinematic calculations of the 6DOF platform;
- A pinhole camera model;
- Kinect data acquisition;
- Visual data processing;
- Calculations of the platform orientation;
- The combination of the 6DOF platform communication and visual data.

NXOpen is a tool which is mainly used for computer-aided engineering (CAE)-purposes, and it allows the user to operate several functions of NX from a centralized location. In this case, NXOpen is used to allow the control of orientation of the platform in all its degrees of freedom, to set the duration of the simulation and to perform a static and dynamic analysis of the motion of the platform. Apart from taking the user input to control its position, it also allows the direct control of the servo angles and it permits the user to import data from the motion of the real-world platform in order to adapt to its orientation. It also starts the LabVIEW application when the simulation method is chosen.

The code behind this graphical user interface (GUI) was written in VB.NET and consists of roughly five parts, with the result given in Figure 5:

- A part that is responsible for maintaining the GUI;
- An object class to do the necessary calculations;
- A method to export the values retrieved from the virtual sensors in the form of a .csv file;
- A component to configure the motion functions of the servomotors;
- A part that regulates the communication with LabVIEW through the use of registry keys.

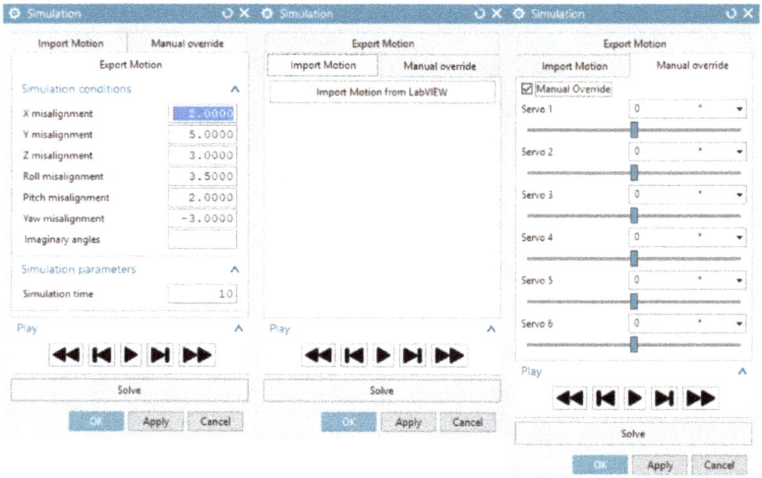

Figure 5. The NXOpen interface with all three tabs visible—here, the user can specify supervisory setpoint values to position the platform before docking.

As observed in Figure 5, the user interface is divided inti three tabs:

- Export motion is used to configure the attitude and position of the virtual platform. The *ids* values of servomotors are written in the text box besides the imaginary angles while filling in the values for the platform orientation—this makes it so that the values can still be adjusted before starting the solution. The simulation time defines the duration of the simulation;
- Import motion: In this mode, a sequence of motion of the real-world-platform is imported and transformed to angles of the servomotors using the inverse kinematics, and then every point is plotted to form a motion-function. This motion function is then used as the driving function for the servo drivers;
- Manual override is used to configure the angles of the individual servomotors. This function is primarily provided for testing purposes and has no influence on the real-world simulator.

The overview of these three operating modes is illustrated in the flowchart in Figure 6, where the red path represents the manual override mode, the blue path represents the import motion mode and the green path represents the export motion mode. The purple dash arrows represent the registry keys (SimStart, PathReady, LVMode, ImportPath) that are used to communicate with the LabVIEW application and ensure that the LabVIEW program does not skip steps the next time it executes. The overview of the working application (Figure 6) shows the main steps in the communication and data processing parts necessary to allow the real-time co-simulation of the real world and virtual world platform setups as part of the docking mechanism.

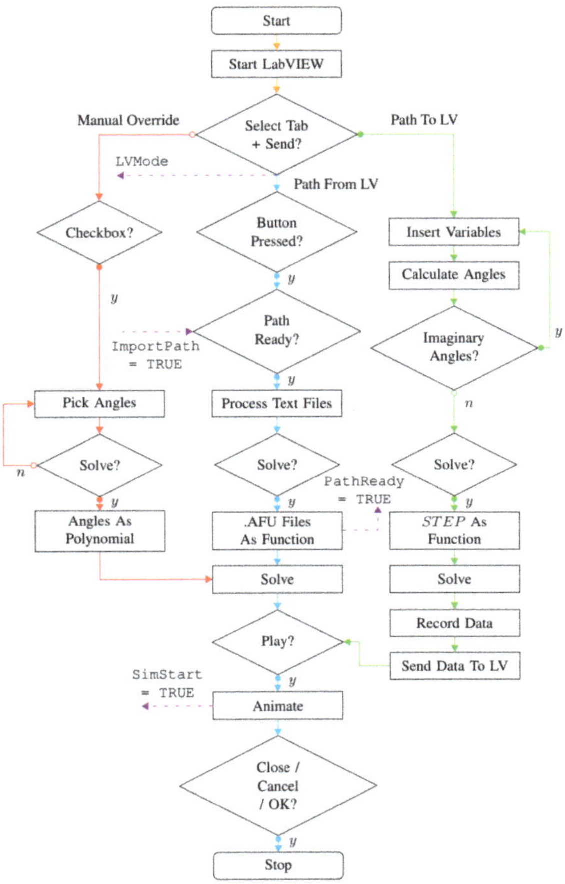

Figure 6. Flowchart of the operational procedures in the VisualBasic application (red path represents the manual override mode, the blue path represents the import motion mode and the green path represents the export motion mode). LV is the abbreviation of LabVIEW.

2.1. Co-Simulation Procedure

The previous section gave a brief overview of both the LabVIEW and the VB.NET applications, while this section explains the cooperation between both applications. Both applications have the ability to influence each other, but only at specific points during the execution of the code, by reading and writing registry keys and by creating intermediary files to transfer large amounts of data. The communication via the registry keys is simply used as a confirmation and allows one application to know which part of the code the other application is executing. It can be said that the communication between LabVIEW and the VB.NET-application is a two-way communication process, but it is however not a simultaneous or continuous type of communication. Larger amounts of data are transferred between applications using both .csv-files and .txt-files, which results in another type of indirect communication. The motion sequence is recorded by six virtual sensors that are positioned at the center of the upper surface of the virtual platform. These sensors measure the displacement in three directions as well as the angular displacement around three axes. NX outputs the measurement data of the sensors as a binary file. These data are used as an input by LabVIEW to match the motion sequence of the real-world platform to that of the virtual platform. Because the data of the virtual sensors are used as a function of the virtual world coordinate system, this data have to be transformed to the real world-platform coordinate system using the principle illustrated in Figure 4b. The real-world-platform motion-sequence is recorded by the Kinect sensor, which returns the platform orientation data. The data consist of three markers on the plate, which are detected by the visual sensor and are further used to calculate the platform position and attitude. The initial locations for each of these points is known, thus enabling us to calculate the transformation matrix between the initial location and the current location.

This co-simulation consists of two types of simulations:

1. The real-world simulation is in itself a real-time simulator: values that are changed on the front panel of the LabVIEW-app have an immediate effect on the orientation of the real-world platform;
2. The virtual-world simulation, however, is a slower-than-real-time-simulation because this application solves the entirety of the sequence at once before starting the animation.

The program mode that most closely resembles a synchronous simulation is the auto-mode. It records small motion-sequences with a step duration of approximately one 60th of a second, sends them to the virtual-world simulator and animates the simulation.

2.2. Simplified Model

The variation in the reflected mass for a single LEMA has been generated from real-life simulators and the data analyzed thoroughly in collaboration with QINETIC Space Kruibeke, Belgium (a confidentiality agreement is in place). To illustrate the high LPV degree in this berthing and docking mechanism system, we introduce a linear model approximation for the extreme cases. Since the leg positions must be controlled, these models capturing the leg dynamics can be approximated in the form indicated bellow:

$$P(s) = \frac{K(s+z_0)(s+z_1)(s+z_1^*)}{(s+p_0)(s+p_1)(s+p_2)(s+p_2^*)} \tag{4}$$

where K is the gain, z_0 is a real zero, z_1^* denotes the complex comprised of z_1, p_0, p_1 real poles and p_2^* denotes the complex comprised of p_2. This form of the transfer function has LPV parameter dynamics; i.e., changes in the z_1 and p_1 values. In order to illustrate the variability of the system, two extreme cases have been considered. The assumed linear model for case 1 is

$$P_1(s) = 2500\frac{(s+625)(s+(1.95+13.4i))(s+(1.95-13.4i))}{(s+53)(s+0.1)(s+(294+334i))(s+(294-334i))} \tag{5}$$

The other extreme (case 2) of LPV dynamics is expressed by the system transfer function:

$$P_2(s) = 2500\frac{(s+628)(s+(5\cdot 10^{-4}+0.2i))(s+(5\cdot 10^{-4}-0.2i))}{(s+53)(s+3\cdot 10^{-5})(s+(292+322i))(s+(292-322i))} \tag{6}$$

In this form of the models, the variations in pole/zero locations are clearly visible.

3. Adaptive Control

Originally, the controller used for this level was a single PI (proportional–integral) controller with feedforward action. An integer PI-cascaded controller has been outperformed in terms of robustness for maintaining closed loop specifications by a fractional order PI controller while being validated on the real-life simulator for European Space Agency at QINETIQ nva Antwerp, Belgium [16]. A robust fractional order PI controller has been proposed and simulated in [6]. Here, we use the adaptive robust fractional order PI (FOPI) control design described in [5]. It has already been shown that fractional order control is a good candidate for the robust control of DC motor components [17–19].

The FOPI described hereafter has the form:

$$C_{FOPI}(s) = K_p(1 + \frac{K_i}{s^{\mu}}) \quad (7)$$

where K_p and K_i are the proportional and integral gains and $\mu \in (0, 1)$ the fractional order of integration. Note that the traditional PI controller is obtained for the special case when $\mu = 1$. The fractional order PI controller in (7) is thus a generalization of the integer order PI controller. The advantages of the fractional order PI controller stem from the extra tuning parameter μ, which can be used to enhance the robustness of the controller. Since there are three tuning parameters in the fractional order PI controller, three performance specifications are used. These three performance criteria lead to the typical tuning rules for fractional order controllers. and they are based on the relationship between time-domain performance specifications and corresponding frequency domain specifications. We enumerate here the most commonly used specifications:

- Gain cross over frequency ω_{gc}, which is related to the settling time of the closed loop system—large values will yield smaller settling times at the cost of higher control efforts;
- Phase margin (PM), which is stability related and an indicator of the closed loop overshoot percentage—usual values are within the 45–65° interval [20];
- The iso-damping property, which is a condition ensuring robustness to gain changes such that overshoot remains constant within a certain gain variation range.

To ensure a constant overshoot, a constant phase margin needs to be maintained around the desired gain crossover frequency, which ultimately implies that the phase of the open-loop system must be kept constant around the specified gain cross-over frequency. In other words, the derivative of the phase with respect to frequency around the gain cross over frequency must be (almost) null. The three performance specifications mentioned above are mathematically expressed as

$$|H_{open-loop}(j\omega_{gc})| = 1 \quad (8)$$

$$\angle H_{open-loop}(j\omega_{gc}) = -\pi + PM \quad (9)$$

$$\frac{d(\angle H_{open-loop}(j\omega))}{d\omega}|_{\omega=\omega_{gc}} \quad (10)$$

where $H_{open-loop}(s) = C_{FOPI}(s)P(s)$ stands for the open loop transfer function and $P(s)$ is the system to be controlled. Equations (8)–(10) require the magnitude, phase and derivative of the phase for both the FOPI controller and $P(s)$. The modulus and phase of the FOPI controller are given by

$$\begin{aligned} |C_{FOPI}(j\omega)| &= K_p\sqrt{1 + 2K_i\omega^{-\mu}\cos\frac{\pi\mu}{2} + K_i^2\omega^{-2\mu}} \\ \angle C_{FOPI} &= -\arctan\left(\frac{K_i\omega^{-\mu}\sin\frac{\pi\mu}{2}}{1+K_i\omega^{-\mu}\cos\frac{\pi\mu}{2}}\right) \end{aligned} \quad (11)$$

To determine analytically the derivative of the phase of the FOPI controller, the phase equation in (11) is used, resulting in

$$\frac{\mu K_i\omega_{gc}^{-\mu-1}\sin\frac{\pi\mu}{2}}{1 + 2K_i\omega_{gc}^{-\mu}\cos\frac{\pi\mu}{2} + K_i^2\omega_{gc}^{-2\mu}} = 0 \quad (12)$$

Equations (8)–(10) also use the magnitude, phase and derivative of the phase of $P(s)$. The values for the modulus, phase and phase slope of the process should be known a priori at the gain cross over frequency ω_{gc}. These can be obtained analytically based on a model of $P(s)$. Assuming a process model is given by Equation (5), the magnitude, phase and derivative of the phase of $P(s)$ are computed as

$$
\begin{aligned}
&|P(j\omega_{gc})| = K \cdot \frac{\sqrt{\omega_{gc}^2+z_0^2}\cdot\sqrt{\omega_{gc}^2+z_1^2}\cdot\sqrt{\omega_{gc}^2+z_1^{*2}}}{\sqrt{\omega_{gc}^2+p_0^2}\cdot\sqrt{\omega_{gc}^2+p_1^2}\cdot\sqrt{\omega_{gc}^2+p_2^2}\cdot\sqrt{\omega_{gc}^2+p_2^{*2}}} \\
&\angle P(j\omega_{gc}) = \tan^{-1}\left(\frac{\omega_{gc}}{z_0}\right) + \tan^{-1}\left(\frac{\omega_{gc}}{z_1}\right) + \tan^{-1}\left(\frac{\omega_{gc}}{z_1^*}\right) - \tan^{-1}\left(\frac{\omega_{gc}}{p_0}\right) - \tan^{-1}\left(\frac{\omega_{gc}}{p_1}\right) \\
&\qquad - \tan^{-1}\left(\frac{\omega_{gc}}{p_2}\right) - \tan^{-1}\left(\frac{\omega_{gc}}{p_2^*}\right) \\
&\frac{d\angle P}{d\omega_{gc}} = \frac{z_0}{\omega_{gc}^2+z_0^2} + \frac{z_1}{\omega_{gc}^2+z_1^2} + \frac{z_1^*}{\omega_{gc}^2+z_1^{*2}} - \frac{p_0}{\omega_{gc}^2+p_0^2} - \frac{p_1}{\omega_{gc}^2+p_1^2} - \frac{p_2}{\omega_{gc}^2+p_2^2} - \frac{p_2^*}{\omega_{gc}^2+p_2^{*2}}
\end{aligned} \tag{13}
$$

To mimic a space rendezvous, the relative mass on each leg of the 6DOF platform can be varied at times during the simulation. In practice, the relative mass can be estimated continuously from the measured position and velocity of the platform. However, a continuous adaptation of controller parameters is not necessary and significantly increases the computational burden of the deployed system operation. Instead, using the standard performance index, a tolerance interval of $\pm 10\%$ is introduced in the overshoot ($OS\% < 20\%$) and settling time ($Ts < 0.1s$) values.

The relations in (13) are easily computed online from previous samples' reflected mass estimation. However, an additional condition of the violation of the tolerance interval is implemented to prevent the sluggish and unnecessary adaptation of controller parameters and consequently discretization and deployment on real-life emulator. The latter steps are performed solely when the threshold of tolerance interval is exceeded. Then, replacing (11) and (12) in (8)–(10) leads to

$$
\begin{aligned}
&K_p\sqrt{1+2K_i\omega_{gc}^{-\mu}\cos\frac{\pi\mu}{2}+K_i^2\omega_{gc}^{-2\mu}} = \frac{1}{|P(j\omega_{gc})|} \\
&\frac{K_i\omega_{gc}^{-\mu}\sin\frac{\pi\mu}{2}}{1+K_i\omega_{gc}^{-\mu}\cos\frac{\pi\mu}{2}} = \tan(\pi - PM + \angle P(j\omega_{gc})) \\
&\frac{\mu K_i\omega_{gc}^{-\mu-1}\sin\frac{\pi\mu}{2}}{1+2K_i\omega_{gc}^{-\mu}\cos\frac{\pi\mu}{2}+K_i^2\omega_{gc}^{-2\mu}} + \frac{d\angle P(j\omega)}{d\omega}|_{\omega=\omega_{gc}} = 0
\end{aligned} \tag{14}
$$

With respect to the digital implementation of this controller, two more parameters are practically relevant: the degree of approximation and the user-specified ω-range in which the desired behavior should be obtained, and tuning rules are given in [21] (includes the Matlab function code provided in the paper).

4. Results and Discussion

The controller parameter solution for the isodamping property is achieved fora gain crossover frequency ω_{gc} of 150 rad/s and phase margin (PM) of 50°. The assumed linear model is (5) and the resulting controller is

$$
C_{FOPI}(s) = 0.065\left(1+\frac{1.15\cdot 10^4}{s^{1.656}}\right) \tag{15}
$$

The parameters of the FOPI controller in (15) were obtained by solving the set of nonlinear equations in (14). First of all, the K_i parameter was estimated as a function of the fractional order μ, using the phase margin and the iso-damping, according to the second and third equations in (7). The two functions were then plotted as indicated in Figure 7a. The intersection of the two functions gives the two control parameters that ensure both a certain phase margin and the iso-damping property. According to Figure 7a, the intersection point yields $K_i = 1.15 * 10^4$ and $\mu = 1.656$. Once these two parameters are computed, the proportional gain is estimated from the magnitude from the (first) equation in (14) as $K_p = 0.065$. Figure 7b presents the Bode diagram for the controller and the two systems denoting two extreme cases (P_1 and P_2) of LPV values (i.e., a variation of 250% in the reflected mass value in the 6DOF platform simulator).

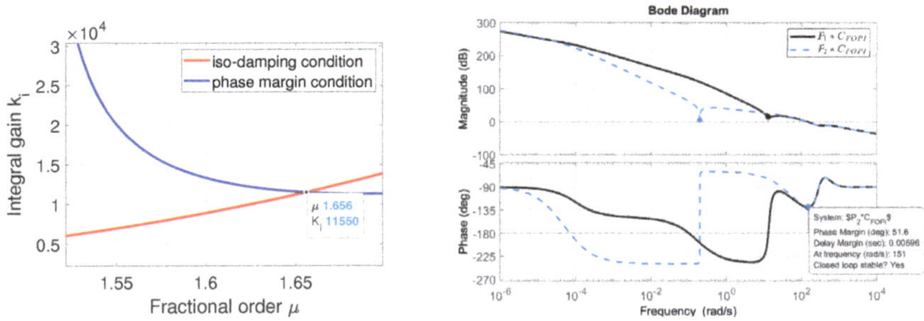

(a) (b)

Figure 7. (**a**) The cross-section point of the phase margin and iso-damping criteria; (**b**) a Bode diagram of the open loop system in the two extreme cases.

The discretization scheme proposed in [21] has been employed to obtain a discretized version of pole–zero pairs. This scheme is particularly efficient at delivering stable, low-order approximations of the non-rational function of the FOPI into a proper filter form, and for our study case, it delivers a fifth order rational transfer function.

An example of a co-simulation interface and real-time tracking between the 6DOF platform and virtual model is illustrated in Figure 8. In this figure, the real world platform can be observed, including the active part of the BDM system (label 1), the front panel of the LabVIEW application (label 2) and the 3D model of the simulated platform (label 3).

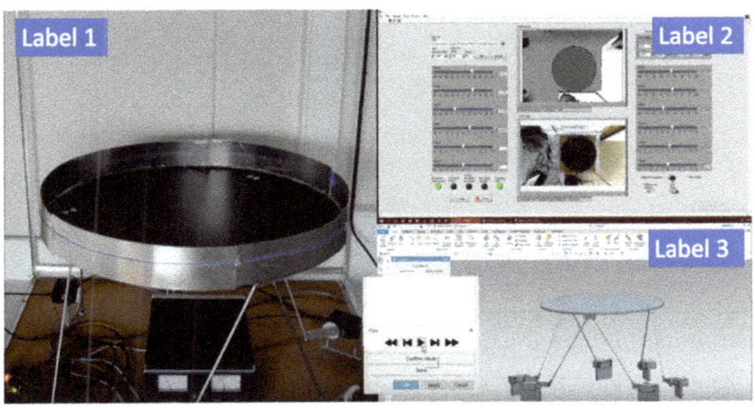

Figure 8. Example of co-simulation interface and real-time tracking between the 6DOF platform and virtual model.

The adaptive controller maintains the same specifications in closed loop performance for the two variations in the reflected mass parameters, with the step response for three platform legs (the other are similar) given in Figure 9. It can be observed that the controller maintains performance despite the great variability in system parameters mimicking the reflected mass for one leg with or without contact at docking.

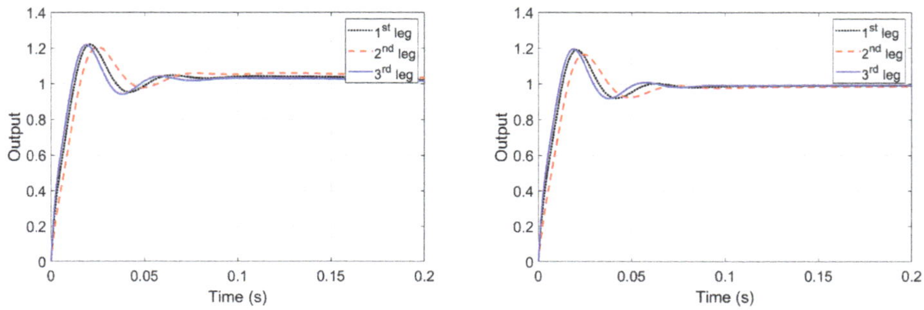

Figure 9. Step responses for the adaptive controller for three legs for the first case, $P_1(s)$ (**left**), and the second case, $P_2(s)$ (**right**).

5. Conclusions

In this paper, we presented a co-simulation framework for a real-world and virtual-world 6DOF platform to mimic the dynamic variability in a docking mechanism for a space rendezvous. The problem of controlling a challenging LPV system using a single type of controller in all operational zones has been addressed. A new approach has been introduced by using fractional order controllers, and a rationale for a generalized approach for LPV systems has been proposed. The simulation results indicate that the proposed approach works satisfactorily and supports the claim that fractional order control is an appealing solution to complex control problems. Further analysis of the control performance in the presence of unknown varying time delays can be evaluated in terms of tele-operation delays for maintenance, failure emergency cases, etc.

Author Contributions: Conceptualization, C.C. and C.M.I.; methodology, C.M.I., C.I.M. and C.C.; software, C.I.M., C.M.I. and C.C.; validation, M.B., C.M.I., C.I.M. and C.C.; formal analysis, C.M.I.; investigation, C.M.I.; resources, C.C., C.M.I. and M.B.; data curation, C.C. and C.M.I.; writing—original draft preparation, C.M.I. and C.C.; writing—review and editing, M.B., C.M.I., C.I.M. and C.C.; visualization, M.B., C.M.I., C.I.M. and C.C.; supervision, C.M.I.; project administration, C.C. and C.M.I.; funding acquisition, C.M.I. All authors have read and agreed to the published version of the manuscript.

Funding: This research received no external funding.

Institutional Review Board Statement: Not applicable.

Informed Consent Statement: Not applicable.

Data Availability Statement: Not applicable.

Acknowledgments: The authors acknowledge Reinder Boven and Nicolas Van Oevelen for their work on this setup during their Masters thesis at Ghent University and Ghent University Special Research Fund MIMOPREC 020-018. Part of this work has been done within a Master thesis program at Ghent University in collaboration with QINETIQ nva Antwerp, Belgium, a subcontractor of European Space Agency. The real parameters of the simulator are not disclosed in this paper in accordance with the confidentiality agreement between the parties. Cristina Muresan was supported by a grant from the Romanian Ministry of Education and Research, CNCS-UEFISCDI, project number PN-III-P1-1.1-TE-2019-0745, within PNCDI III.

Conflicts of Interest: The authors declare no conflict of interest. The funders had no role in the design of the study; in the collection, analyses, or interpretation of data; in the writing of the manuscript, or in the decision to publish the results.

References

1. Kwon, S.; Shimomura, T.; Okubo, H. Pointing control of spacecraft using two SGCMGs via LPV control theory. *Acta Astronaut.* **2011**, *68*, 1168–1175. [CrossRef]
2. Saleh, J.; Hastings, D.; Newman, D. Flexibility in system design and implications for aerospace systems. *Acta Astronaut.* **2003**, *53*, 927–944. [CrossRef]
3. Mankins, J. Technology readiness assessments: A retrospective. *Acta Astronaut.* **2009**, *65*, 1216–1223. [CrossRef]
4. Vilanova, R.; Visioli, A. *PID Control in the Third Millennium*; Advances in Industrial Control; Springer: London, UK, 2012.
5. Muresan, C.I.; Copot, C.; Ionescu, C.M. Robust fractional order control of LPV dynamic mechatronic systems. In Proceedings of the IEEE 15th Int Conf on Control and Automation (ICCA), Edinburgh, UK, 16–19 July 2019; IEEE: Edinburgh, UK, 2019; pp. 154–159.
6. Ionescu, C.M.; Van Oevelen, N.; Goharoodi, S.K.; Paijmans, B.; De Keyser, R. Control of LPV Mechatronic Systems in Presence of Dynamic Uncertainties. In Proceedings of the 2017 IEEE 21st International Conference on Intelligent Engineering Systems (INES), Larnaca, Cyprus, 20–23 October 2017; IEEE: Larnaca, Cyprus, 2017; 6p.
7. Padula, F.; Visioli, A. *Advances in Robust Fractional Control*; Advances in Industrial Control; Springer: London, UK, 2015.
8. Preumont, A.; Horodinca, M.; Romanescu, I.; de Marneffe, B.; Avraam, M.; Deraemaeker, A.; Bossens, F.; Hanieh, A.A. A six-axis single-stage active vibration isolator based on Stewart platform. *J. Sound Vib.* **2007**, *300*, 644–661. [CrossRef]
9. Yao, R.; Fu, C.; Sun, C.; Zhu, W. Accuracy design of the Stewart manipulator of the Five-hundred-meter Aperture Spherical radio Telescope. *Adv. Mech. Eng.* **2019**, *11*, 1–9. [CrossRef]
10. Zhong, Y.; Dutta, A.; Copot, C.; Ionescu, C.M.; De Keyser, R. Implementation of a fractional PD controller tuned by genetic algorithm for a Stewart platform. In Proceedings of the 2013 9th Asian Control Conference (ASCC), Istanbul, Turkey, 23–26 June 2013; IEEE: Istanbul, Turkey, 2013; pp. 1–6; doi:10.1109/ASCC.2013.6606245. [CrossRef]
11. Copot, C.; Ionescu, C.M.; De Keyser, R. Visual servo control of a Stewart platform using fractional-order PID controller. In Proceedings of the 2014 18th International Conference on System Theory, Control and Computing (ICSTCC), Sinaia, Romania, 17–19 October 2014; IEEE: Sinaia, Romania, 2014; pp. 70–75.
12. Copot, C.; Zhong, Y.; Ionescu, C.M.; De Keyser, R. Tuning fractional PID controllers for a Stewart platform based on frequency domain and artificial intelligence methods. *Cent. Eur. J. Phys.* **2013**, *11*, 702–713.
13. Chevalier, A.; Copot, C.; Ionescu, C.M.; De Keyser, R. Automatic calibration with robust control of a six DoF mechatronic system. *Mechatronics* **2016**, *35*, 102–108. [CrossRef]
14. Copot, C.; Ionescu, C.; Muresan, C. *Image-Based and Fractional-Order Control for Mechatronic Systems*; Advances in Industrial Control; Springer: London, UK, 2020.
15. Microsoft, Coordinate Mapping. 2014. Available online: https://docs.microsoft.com/en-us/previous-versions/windows/kinect/dn785530(v=ieb.10) (accessed on 19 April 2021).
16. Paijmans, B.; Gracia, O.; Dittmer, H.; Meuws, D.; Van Oevelen, N. International Berthing Docking Mechanism Model-based Control design and Verification of a Stewart platform. In Proceedings of the 10th International ESA Conference on Guidance, Navigation & Control Systems, Salzburg, Germany, 29 May–2 June 2017.
17. Ionescu, C.; Dulf, E.; Ghita, M.; Muresan, C. Robust Controller Design: Recent Emerging Concepts for Control of Mechatronic Systems. *J. Frankl. Inst.* **2020**. [CrossRef]
18. Petras, I. *Fractional Order Nonlinear Systems*; Nonlinear Physical Science; Springer: London, UK, 2011.
19. Hosseinnia, S.; Tejado, I.; Vinagre, B. Fractional order reset control—Application to a servomotor. *Mechatronics* **2013**, *23*, 781–788. [CrossRef]
20. Ionescu, C.M.; De Keyser, R. The next generation of relay based PID autotuners (PART1): Some insights on the performance of simple relay based PID autotuners. In *IFAC Conf on Advances in PID Control*; IFACPapersOnline: Brescia, Italy, 2012; pp. 122–127.
21. De Keyser, R.; Muresan, C.I.; Ionescu, C.M. An efficient algorithm for low-order direct discrete-time implementation of fractional order transfer functions. *ISA Trans.* **2018**, *74*, 229–238. [CrossRef] [PubMed]

applied sciences

MDPI

Article

State and Parameter Estimation of a Mathematical Carcinoma Model under Chemotherapeutic Treatment

Máté Siket, György Eigner *, Dániel András Drexler, Imre Rudas and Levente Kovács

Physiological Research Control Center, Óbuda University, Bécsi Street 96/B, H-1034 Budapest, Hungary; siket.mate@nik.uni-obuda.hu (M.S.); drexler.daniel@nik.uni-obuda.hu (D.A.D.); rudas@uni-obuda.hu (I.R.); kovacs@uni-obuda.hu (L.K.)

* Correspondence: eigner.gyorgy@nik.uni-obuda.hu; Tel.: +36-30-625-1120

Received: 6 November 2020; Accepted: 13 December 2020; Published: 17 December 2020

check for updates

Abstract: One challenging aspect of therapy optimization and application of control algorithms in the field of tumor growth modeling is the limited number of measurable physiological signals—state variables—and the knowledge of model parameters. A possible solution to provide such information is the application of observer or state estimator. One of the most widely applied estimators for nonlinear problems is the extended Kalman filter (EKF). In this study, a moving horizon estimation (MHE)-based observer is developed and compared to an optimized EKF. The observers utilize a third-order tumor growth model. The performance of the observers is tested on measurements gathered from a laboratory mice trial using chemotherapeutic drug. The proposed MHE is designed to be suitable for closed-loop applications and yields simultaneous state and parameter estimation.

Keywords: moving horizon estimation; Kalman filter; parameter estimation; tumor growth estimation

---

1. Introduction

Cancer treatment and its related fields are intensively studied subjects, since cancer is a major cause of death globally. It is estimated that there will be 21.4 million cases annually by 2030 [1,2]. The application of modern control algorithms for cancer treatment has much potential—especially in the adjustment of the dosages—however, there are difficulties as well [3]. The article investigates a facet of a larger project. The larger project aims to minimize the amount of administered chemotherapeutic drug—while retaining its effect—by applying personalized treatments. Lower amount of doses expected to reduce side effects and potentially postpone the emergence of tumor resistance. One approach is the utilization of feedback control, where the controller (which decides about the drug administration scheme) adapts to the patient. Based on mathematical models accurately describing the underlying physiological processes the professionals have the possibility to define appropriate therapies. System engineers thrive to provide a proper description of the tumor growth, however, the inter- and intrapatient variability and the effects of different drugs are cumbersome to model. Several models were introduced over the years [4–9]. In particular, we are concerned with the use of a third-order model. The model describes the dynamics of the living and dead tumor cells, as well as drug concentration [7]. The output of the system is the total tumor volume (sum of the living and the dead tumor volumes).

In tumor growth modeling there are variables that cannot be measured, or are not feasible to measure in practice on a day-to-day basis. Such variables in this particular case are the individual volume of the living and dead tumor cells and drug concentration or the model parameters characterizing the patient variability. A mathematical model gives the general behavior of the system. The general structure can be tailored to each individual by the model parameters. For instance, there is a parameter that corresponds to the clearence of the drug, which can vary from patient to patient. Although they cannot be measured, they play an important role in describing the dynamics of the system and are often required for specific therapy optimization or closed-loop algorithms. In order to extract those unmeasurable variables, usually, an observer is applied. The goal of an observer is to provide information about the inner state variables or parameters based on the measurements. The most common algorithm is the Kalman filter (KF) [10], however, other approaches like sliding mode observer, particle filter [11], linear observer [12] or moving horizon estimation (MHE) [13] are also employed. The latter one is an optimization-based approach, where the name suggests that the optimization is performed on a moving window. The window makes it possible to take into account a series of measurements, and the optimization makes it possible to define application-specific constraints and penalizations. In this paper, an MHE is developed and its performance is compared to an optimized extended Kalman filter (EKF). Often, the drawback of the studies, e.g., [10–13] is that they are performed on simulated data which is favorable from several aspects. In contrast, this study is evaluated on laboratory measurements, where noise, irregular sampling times and most importantly, structural mismatch are present.

The paper is structured as follows. Section 2.2 describes the utilized third-order tumor growth model and previously identified parameter sets. In Sections 2.3 and 2.4 first theoretical, general description for the EKF and MHE based observers are given. Thereafter the observers are applied to the specific tumor growth model. Measurement noise is characterized in Section 2.3.1. Sections describing the observers end with the tuning methods. In Section 3 the results are summarized. The observers are evaluated based on their capabilities following the trends of the measurements. During the assessment, the fluctuations of the parameters are taken into account too. Furthermore, different window lengths of the MHE are investigated as well. Finally, in Sections 4 and 5 the paper is concluded.

2. Materials and Methods

2.1. Experimental Data

The experiments were conducted by the Membrane Protein Research Group of the Research Centre for Natural Sciences published in [14]. The mice are identified with the "PLD" acronym, indicating that the injected chemotherapeutic drug was the pegylated liposomal doxorubicin (PLD) during the experiments. In this trial tumor pieces from a Brca1-/-;p53-/- mouse tumor were transplanted orthotopically to syngenic FVB mice to reduce the time of tumor formation. The tumor volume was measured by calipers every 1–5 days. The treatment protocol was a dynamic reactive approach which was defined as follows. The mice were treated based on the decision of the professionals. The injection could be granted if the tumor volume reached 200 [mm^3] and at least 10 days past since the last treatment. The injected drug was the maximum tolerable dose 8 [mg/kg] every time. The experiment focused on the elimination of the primary tumor. The experiment was stopped when the tumor volume grew above the 2000 [mm^3] threshold.

In Figure 1, an example is given of the trial. The raw caliper measurements are represented with black circles, the treatment threshold is plotted with a dashed grey line and the solid green line is a linear interpolation between the measurements. It can be seen that the particular mouse responded well to the treatment, since the volume decreased after the injections. However, some mice develop resistance and the drug stops taking effect (meaning continuously growing volume regardless of the presence of the drug). Those who developed resistance are not taken into account in this study as the applied model does not consider resistance. As a result, data from mice PLD1, PLD2 and PLD8 from [14] are not used in this study.

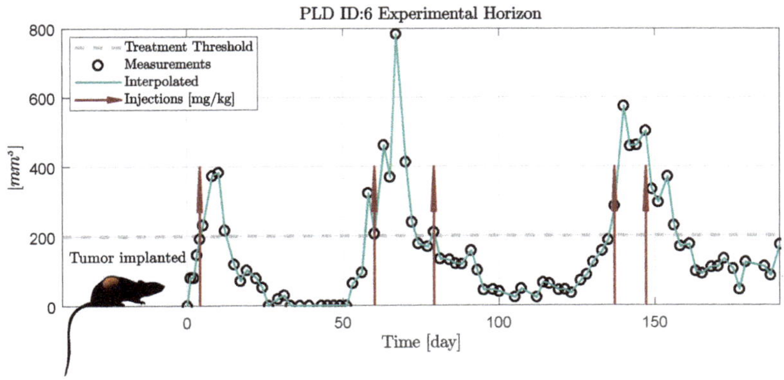

Figure 1. Scheme of the experiment from [14].

2.2. The Applied Tumor Growth Model

The investigated tumor growth model is a third-order nonlinear ordinary differential equation. The model differentiates between living and dead tumor cell volumes by introducing separate state variables: x_1 [mm^3] and x_2 [mm^3], respectively. The third variable x_3 [mg/kg] describes the drug concentration in the host. The dynamics is described by the equations given in [6,7] at time t as

$$\dot{x}_1(t) = (a-n)x_1(t) - b\frac{x_1(t)x_3(t)}{ED_{50}+x_3(t)}, \tag{1}$$

$$\dot{x}_2(t) = nx_1(t) + b\frac{x_1(t)x_3(t)}{ED_{50}+x_3(t)} - wx_2(t), \tag{2}$$

$$\dot{x}_3(t) = -c\frac{x_3(t)}{K_B+x_3(t)} - b_k\frac{x_1(t)x_3(t)}{ED_{50}+x_3(t)} + u(t), \tag{3}$$

where a describes the proliferation rate of the living tumor cells, n is the tumor cell necrosis rate, w is the dead tumor cell washout rate, c is the drug depletion rate. The pharmacodynamics is affected by the b and ED_{50} (median effective dose) parameters. The effect of the drug and the depletion of the drug is characterized by equations following the dynamics of Michaelis–Menten kinetics, thus ED_{50} and K_B are Michelis–Menten parameters. Resistance is not modeled explicitly, but the ineffectiveness of the drug appears as a specific combination of the parameters. When the difference $a-b-n>0$, the drug is not able to shrink the volume, it only decreases the growth rate. Qualitative analysis of the model has been previously done in [15]. The values of the parameters for the mice identified based on a mixed-effect model [7] are given in Table 1.

Table 1. The identified parameters of the mice from experiment specified in [14] using a mixed-effect model [7], along with the mean values and standart deiations for each parameter.

Parameter	PLD1	PLD2	PLD3	PLD4	PLD5	PLD6	PLD8	PLD9	Nominal	SD
a [1/day]	0.333	0.307	0.307	0.310	0.289	0.299	0.308	0.311	0.306	0.0186
b [1/day]	0.116	0.169	0.198	0.180	0.163	0.184	0.174	0.167	0.166	0.0302
c [1/day]	0.235	0.297	0.304	0.272	0.312	0.365	0.187	0.161	0.257	0.0820
n [1/day]	0.115	0.148	0.153	0.173	0.134	0.161	0.133	0.145	0.144	0.0235
$b_k\left[\frac{10^{-7}\text{mg}}{\text{kg}\cdot\text{day}\cdot\text{mm}^3}\right]$	6.15	6.05	6.02	6.10	6.19	6.16	6.17	6.11	6.12	0.404
K_B [mg/kg]	0.367	0.361	0.342	0.230	0.362	0.374	0.515	0.400	0.36	0.1242
ED_{50} [10^{-5}mg/kg]	8.89	9.03	10.4	13.3	8.64	7.91	7.79	8.94	9.71	1.48
w [1/day]	0.346	0.344	0.331	0.341	0.341	0.339	0.336	0.342	0.34	0.0253

2.3. Extended Kalman Filter

For nonlinear systems one of the most widely applied observer is the EKF. First, assume a general system given in the form of:

$$\dot{\mathbf{x}}(t) = f(\mathbf{x}(t), \mathbf{u}(t)) + \mathbf{w}(t), \tag{4}$$

$$\mathbf{y}(t) = h(\mathbf{x}(t), \mathbf{u}(t)) + \mathbf{v}(t), \tag{5}$$

where f describes the nonlinear system dynamics, h is the output equation, $\mathbf{w}$ is the additive process noise or disturbance and $\mathbf{v}$ is the measurement noise. Equation (4) is generalized further with the augmentation of system parameters $\mathbf{x} := [\mathbf{x}_s\ \mathbf{x}_p]^\top$ *s.t.* $\dot{\mathbf{x}}_p = \mathbf{0}$, where $\mathbf{x}_s$ is the original state vector and $\mathbf{x}_p$ is the parameter vector.

Since the measurements are done only at discrete times, continuous-discrete EKF has been applied. In the prediction step the model and covariances are propagated using the forward Euler method with shorter sampling times. The equations of the prediction step in time instant t are defined as

$$\dot{\mathbf{x}}(t) = f(\mathbf{x}(t), \mathbf{u}(t)), \tag{6}$$

$$\mathbf{y}(t) = h(\mathbf{x}(t), \mathbf{u}(t)), \tag{7}$$

$$\dot{\mathbf{P}}(t) = \mathbf{F}(t)\mathbf{P}(t) + \mathbf{P}(t)\mathbf{F}^\top(t) + \mathbf{Q}, \tag{8}$$

$$\mathbf{F}(t) = \left.\frac{\partial f}{\partial \mathbf{x}}\right|_{\mathbf{x}(t),\mathbf{u}(t)} \tag{9}$$

where $\mathbf{P}$ is the error covariance matrix, $\mathbf{F}$ is the Jacobian matrix of the system and $\mathbf{Q}$ is the process noise covariance matrix. The update step in the k-th discrete step is defined by

$$\mathbf{K}[k] = \mathbf{P}[k']\mathbf{H}^\top[k]\left(\mathbf{H}[k]\mathbf{P}[k']\mathbf{H}^\top[k] + \mathbf{R}\right)^{-1}, \tag{10}$$

$$\mathbf{x}[k] = \mathbf{x}[k'] + \mathbf{K}[k]\left(\mathbf{z}[k] - h(\mathbf{x}[k'])\right), \tag{11}$$

$$\mathbf{P}[k] = \left(\mathbf{I} - \mathbf{K}[k]\mathbf{H}[k]\right)\mathbf{P}[k'], \tag{12}$$

$$\mathbf{H}[k] = \left.\frac{\partial h}{\partial \mathbf{x}}\right|_{\mathbf{x}[k']} \tag{13}$$

where $\mathbf{K}$ is the Kalman gain, $\mathbf{H}$ is the Jacobian of the h output equation, $\mathbf{z}$ is the measurement, $\mathbf{R}$ is the measurement noise covariance matrix and the aposthrope k' indicates the a priori values of the update step. The transition from the continuous time domain to the discrete is given by: $\mathbf{x}[k'] = \mathbf{x}(t_k)$.

The EKF structure has been implemented using the model (1)–(3), extended with parameter estimation. The estimated parameters were selected in accordance with a previous sensitivity and identifiability analysis [16]. The results of the analysis indicate that the most sensitive parameters are a, b, n and$w$, thus $\mathbf{x}_p := [a, b, n, w]^\top$. The identifiability analysis with the subset of parameters resulted in a structurally locally identifiable [17] system. This indicates that there is a finite number of parameter sets producing the same trajectories. For the developments of the observers the following assumption is used: $\mathbf{w} = [\mathbf{w}_s\ \mathbf{w}_p]^\top$ *s.t.* $\mathbf{w}_s = \mathbf{0}$. Practically speaking, it is assumed that the mismatch between the measurements and the model rises only from measurement noise and parameter uncertainty. With the augmented state vector the Jacobians are given as:

$$\mathbf{F} = \begin{bmatrix} x_4 - x_5 - \frac{x_6 x_3}{ED_{50}+x_3} & 0 & -x_6 \frac{x_1 ED_{50}}{(ED_{50}+x_3)^2} & x_1 & -x_1 & -\frac{x_1 x_3}{ED_{50}+x_3} & 0 \\ x_5 + \frac{x_6 x_3}{ED_{50}+x_3} & x_7 & x_6 \frac{x_1 ED_{50}}{(ED_{50}+x_3)^2} & 0 & x_1 & \frac{x_1 x_3}{ED_{50}+x_3} & -x_2 \\ -b_k \frac{x_3}{ED_{50}+x_3} & 0 & -c \frac{K_B}{(K_B+x_3)^2} - \frac{b_k x_1 ED_{50}}{(ED_{50}+x_3)^2} & 0 & 0 & 0 & 0 \\ 0 & & \cdots & & & & 0 \\ \vdots & & \ddots & & & & \vdots \\ 0 & & \cdots & & & & 0 \end{bmatrix}, \tag{14}$$

$$\mathbf{H} = \begin{bmatrix} 1 & 1 & 0 & 0 & 0 & 0 & 0 \end{bmatrix}. \tag{15}$$

2.3.1. Measurement Noise Characteristics

The noise characteristics of the measurements are quantified by calculating the standard deviation of the difference between the raw and filtered measurements. The filtered values are calculated by applying a second-order low-pass Butterworth filter in zero-phase setting. The resulting standard deviation $\sigma = 31.39$ is used for the $\mathbf{R} = \sigma^2$ measurement variance. In Figure 2 the measured and filtered pairs are shown on a logarithmic scale to better visualize the deviations at small volumes. Zero measured values are shifted to 0.1 [mm^3], which was chosen arbitrarily for visualizing reasons.

The tumor volumes are measured using calipers and the actual volumes are approximated. This method is less accurate and reliable when smaller volumes are being measured. Data of measurements under 50 [mm^3] are rather scarce too, usually from a measured zero volume at the next sampling, the volume jumps instantly to the proximity of 50 [mm^3].

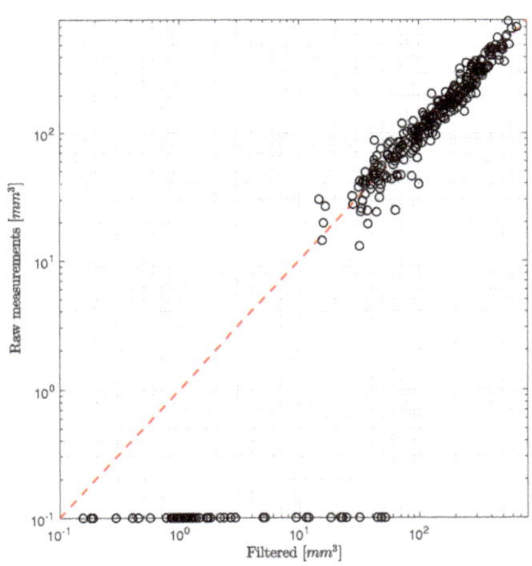

Figure 2. Difference between the raw and filtered measurements.

2.3.2. Kalman Filter Tuning

One critical aspect of comparing the performance of different observers is the tuning. In order to extract the most performance, the elements of the $\mathbf{Q}$ and $\mathbf{P}[0]$ covariance matrix were determined during an optimization. The goal of the optimization was to minimize the root-mean-square error (RMSE) between the measured trajectories and the predicted ones:

$$\min_{\mathbf{Q},\mathbf{P}[0]} \sum_{i=1}^{N} RMSE(\mathbf{z}^{(i)} - \mathbf{y}^{(i)}), \tag{16}$$

where N is the number of treated mice, $\mathbf{z}$ is the vector of measured values of the i-th mouse and $\mathbf{y}$ is the output of the system. The optimized parameters were in the main diagonal corresponding to the estimated parameters and the first element of the $\mathbf{P}[0]$ matrix to guarantee quicker convergence. The optimization is done by the *fmincon* function of the MATLAB environment from 100 initial points generated by the Latin Hypercube Sampling [18]. The optimized matrices are given as:

$$\mathbf{Q} = \begin{bmatrix} 0 & 0 & 0 & 0 & 0 & 0 & 0 \\ 0 & 0 & 0 & 0 & 0 & 0 & 0 \\ 0 & 0 & 0 & 0 & 0 & 0 & 0 \\ 0 & 0 & 0 & 2.976 \times 10^{-6} & 0 & 0 & 0 \\ 0 & 0 & 0 & 0 & 4.022 \times 10^{-5} & 0 & 0 \\ 0 & 0 & 0 & 0 & 0 & 1.103 \times 10^{-9} & 0 \\ 0 & 0 & 0 & 0 & 0 & 0 & 1.437 \times 10^{-3} \end{bmatrix} \tag{17}$$

$$\mathbf{P}[0] = \begin{bmatrix} 2.125 \cdot 10^{3} & 0 & 0 & 0 & 0 & 0 & 0 \\ 0 & 0 & 0 & 0 & 0 & 0 & 0 \\ 0 & 0 & 0 & 0 & 0 & 0 & 0 \\ 0 & 0 & 0 & 4.946 \times 10^{-8} & 0 & 0 & 0 \\ 0 & 0 & 0 & 0 & 1.502 \times 10^{-5} & 0 & 0 \\ 0 & 0 & 0 & 0 & 0 & 8.990 \times 10^{-4} & 0 \\ 0 & 0 & 0 & 0 & 0 & 0 & 2.755 \times 10^{-3} \end{bmatrix} \tag{18}$$

2.4. Moving Horizon Estimation

The MHE is an optimization-based method, where the minimization of a developed cost function is done on a moving window. A moving window of one step can coincide with the widely applied EKF. However, using a longer window and introducing some application-specific terms or constraints can be beneficial in several cases. Taking into account more measurements can result in an estimator more robust to external disturbances or delays.

The cost function can be formulated in several ways, depending on the application. General arguments and terms are summarized in Table 2, where the notations are given according to (4) and (5). The $\mathbf{W}_i$ matrices provide weighting between the terms of the cost function, their subscripts indicates the analogy between the EKF and MHE. The weighting matrices are analogous to the inverses of the covariance matrices (8), (10). The $\mathbf{x}_{ol}$ vector function is the open-loop trajectory from the initial state of the current window, the arrival cost is calculated in the $k{-}M$-th step, where M is the length of the horizon. The difference between the current parameters and the previous window is denoted by $\Delta\mathbf{p}$. Following the assumption that the mismatch between the model and measurements are caused by parameter uncertainty, intrapatient variability and measurement noise, the parameters $\mathbf{p} = [a, b, n, w]$ are optimized, similarly to the EKF design. Given the arguments, the MHE problem can be generarly formulated as follows:

$$\min_{\mathbf{x}[k-m],\mathbf{p}} ||\hat{\mathbf{x}}[k-m] - \mathbf{x}[k-m]||^2_{\mathbf{W}_P} + \sum_{i=k-M}^{k} ||\mathbf{z} - h(\mathbf{x}(\mathbf{p}))||^2_{\mathbf{W}_R} + ||\Delta\mathbf{p}||^2_{\mathbf{W}_Q}. \tag{19}$$

Table 2. Moving horizon estimation (MHE) cost function formulations.

Penalization	Arguments State	Disturbance	Parameter
Measurement difference	$\|\|\mathbf{z} - h(\mathbf{x})\|\|^2_{\mathbf{W}_R}$	$\|\|\mathbf{z} - h(\mathbf{x}(\mathbf{w}))\|\|^2_{\mathbf{W}_R}$	$\|\|\mathbf{z} - h(\mathbf{x}(\mathbf{p}))\|\|^2_{\mathbf{W}_R}$
Modification	$\|\|\mathbf{x} - \mathbf{x}_{ol}\|\|^2_{\mathbf{W}_Q}$	$\|\|\mathbf{w}\|\|^2_{\mathbf{W}_Q}$	$\|\|\Delta\mathbf{p}\|\|^2_{\mathbf{W}_Q}$
Arrival cost	$\|\|\hat{\mathbf{x}} - \mathbf{x}\|\|^2_{\mathbf{W}_P}$	$\|\|\hat{\mathbf{x}} - \mathbf{x}\|\|^2_{\mathbf{W}_P}$	$\|\|\hat{\mathbf{x}} - \mathbf{x}\|\|^2_{\mathbf{W}_P}$

Compared to the general formulation, three modifications are implemented. First, as a worst case scenario the initial state of the horizon is set to the last estimated state. Secondly, the differences are normalized in order to make the weighting between the measurements and modifications easier, scaling the errors to the same order of magnitude. Division with zero is avoided: when the tumor volumes could not be exactly measured due to their small volumes, 1 [mm^3] is assigned to their measured value. Unit volume ensures numerical stability and is physiologically more relevant as the tumors are not completely eliminated. The characteristics of the model is in agreement with this, as in ideal condition the volume converges to zero, but it is bound to be positive. Thirdly, a rational function (23) was introduced, which penalizes the difference less, when the measured tumor volumes are small. Justification for the second and third point is more detialed in Section 2.3.1. Considering these modifications the cost function in the k-th step is defined as

$$\min_{\mathbf{p}} \; J[k] = \sum_{i=k-M}^{k} \left|\left| \frac{z[i]-y[i]}{z[i]} \right|\right|^2_{\mathbf{W}_R[i]} + \left|\left| \frac{\Delta \mathbf{p}}{\mathbf{p}[k-1]} \right|\right|^2_{\mathbf{W}_Q[k]}, \tag{20}$$

$$\text{subject to } \mathbf{W}_R[i] = \frac{1}{M}\omega(i), \tag{21}$$

$$\mathbf{W}_Q[k] = \frac{1}{3T_s[k]} diag(0,1,1,1), \tag{22}$$

$$\omega(i) = \left(\frac{z[i]}{50+z[i]}\right)^2, \tag{23}$$

$$\mathbf{p} \in [\mathbf{lb}, \mathbf{ub}], \tag{24}$$

where T_s is the sampling time of the measurements which is irregular; it varies between 1 and 5 days. This means that the sliding window shall also be irregular, so that when the new measurement is made, the window is adjusted to the time interval which is the closest to the predefined window length. Since the parameters are only modified when new measurements are made, the penalization of the change in parameter values can be different from time to time. In order to counteract this phenomenon, the parameter change penalization is normalized with the last sampling time. The upper and lower bounds for the parameters are defined as $\mathbf{lb} := [0.01, 0.2, 0.1, 0.1]$ and $\mathbf{ub} := [2, 1, 1, 1]$. The bounds were introduced to be large to avoid saturation of variables during the simulation. According to our previous investigations, all four parameters remain in these regions during the model fitting regardless the measured volumes.

For the minimization of the cost function, the inbuilt $fmincon$ function of the MATLAB software was applied. In each step, the initial values for the optimization were the estimates of the previous step which is often called hot start. In the first step, the initial values of the parameters are set to the nominal identified values from Table 1. Furthermore, the implementation of the cost function is extended with stopping criteria, rendering the optimization quicker or feasible at all. The intrinsic property of the model is that the tumor can grow even in cases when it is under treatment. It can occur that the optimizer selects parameter combinations which results in unrealistically large tumor volumes, thus making the solution of the differential equations unstable. To avoid such situations the propagation of such trajectories are stopped and penalized. The stopping threshold is based on clinical protocol, since the experiment has to be stopped, when the tumor volume grows larger 2000 [mm^3]. The cost is defined to be dependent on the volume, since otherwise assigning a constant value or infinity can make the optimizer stuck due to constant gradients of the object function. The change of parameter a is not penalized as it can be seen in (22). The reason behind this is that the tumor growth is dependent on the $ax_1(t)$ term. In the case of small tumor volumes, the model could not describe quickly varying tumor growth, unless more sudden changes are allowed for the a parameter. In the case of non-estimated parameters the nominal values were assigned from Table 1.

3. Results

The observers were evaluated on the data described in Section 2.1. As mentioned each mouse is labeled with PLD followed by an ID number. Estimation horizons of the MHE were tested in the 7–35 days range, however for better visibility only the 14 day version is shown beside the EKF.

Table 3 compares the RMSE of the measured and estimated volumes between the EKF and the investigated MHEs, given for each investigated mouse. When comparing the results two points must be emphasized. First, the assumptions made during the design of the observers. Secondly, the worst case scenario is investigated, in the sense that the EKF was tuned based on an optimization, where all the mice were taken into account. On the other hand, the weights of the MHE was not optimized only a penalizing function was introduced and applied instead of other tuning process. The RMSE indicates that for horizons shorter than 20 days the MHE can achieve better approximations on average. At horizon length of 28 days the results are varying from patient to patient. Only at 35 days the EKF has a clear edge over the MHE. Increasing the horizon of the MHE deteriorated the approximations as the actual parameter set had to describe longer sections, where structural mismatch and/or parameter variation occurs due to the change in the physiology of the tumor. At this point, it has to be emphasized again that while the MHE takes into account a series of measurements, the EKF works from sample to sample, thus various horizon lengths cannot be adapted.

Table 3. RMSE between the measured and estimated volumes.

	EKF	MHE 7 Days	MHE 14 Days	MHE 21 Days	MHE 28 Days	MHE 35 Days
PLD3	121.6	114.6	115.5	121.5	121.4	169.5
PLD4	72.12	32.82	48.96	59.33	60.33	103.9
PLD5	74.53	28.68	40.48	52.16	61.89	88.76
PLD6	97.93	43.46	50.15	65.97	88.49	146.1
PLD9	59.55	49.05	55.67	61.77	72.52	86.39
MEAN	85.14	53.73	62.16	72.16	80.94	118.9

Figure 3 depicts the comparison of the time horizons using different measures. Three measures, namely mean absolute error (MAE), "noise" and standard deviation (SD) of each estimated parameter were selected in order to assess the performance of the observers and horizons. The MAE describes the general accuracy with respect to the measurements, the "noise" term quantifies the smoothness of the estimation by calculating the number of direction changes along the trajectory. The SD of each estimated parameter gives information about the range in which the parameter had to vary in order to provide an optimized fit. One can inspect that the MAE increases in a quasi-linear way with respect to the horizon. The smoothest trajectory is achieved by using 14-day-long horizon and the SD of parameter a is the lowest around 21 days. Concerning the penalized parameters b, w, n the lowest SD is achieved by using the 7-day-long horizon. The main difference between the EKF and MHE is that the optimization of the EKF resulted in very small gains for parameters a and n. This is anticipated as those parameters directly affect the tumor growth, with larger gains it may be possible to achieve better approximation for one patient, but it could lead to bigger divergence for a population. The divergence is avoided by the MHE as the optimization is constrained, parameter sets producing trajectories that reach 2000 [mm^3] are neglected.

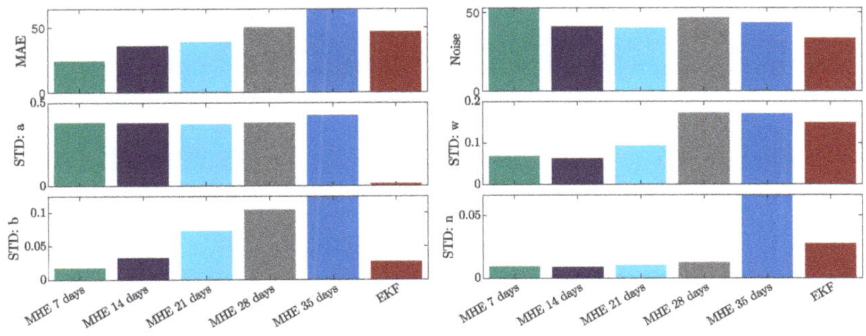

Figure 3. Comparison of the investigated observers.

In Figure 4, an example (PLD ID:5) is given about the effect of the rational penalizing function. The function was introduced in (21). The two estimated trajectories, denoted by *with* $\omega(\cdot)$ and *without* $\omega(\cdot)$ are the output of the MHE with 14 days of horizon length. For visualization only the 14 day scenario is given, however, the effect is similar in the other cases, and the delay is more pronounced towards the longer horizons. It can be seen that without this weighting in the case of small volumes there is a significant delay in the response of the observer. The dashed purple lines show the value of the function with respect to time ($\omega(t)$, left subfigure) and the measured volume ($\omega(x)$, right subfigure).

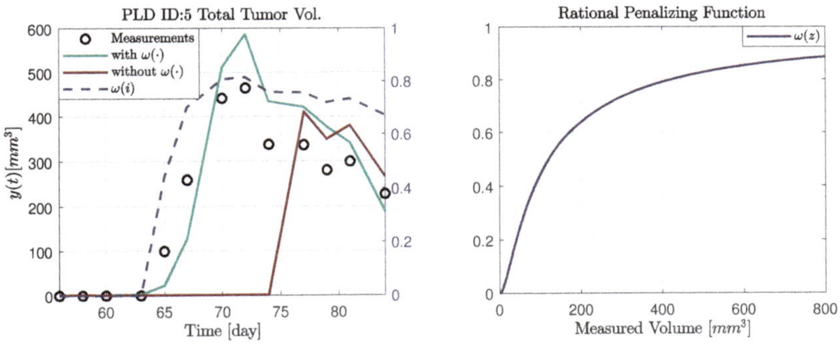

Figure 4. Effect of the rational penalizing function.

In the following explanatory examples we address the problem of intrapatient variability, and how the observers were able to overcome it. Figures 5–14. showcase the state and parameter estimations separately. The black circles denote the measurements, however, it is important to note that only the total tumor volume is measured, i.e., $y = x_1 + x_2$. The dashed magenta lines are the results from a Stochastic Approximation of Expectation-Maximization identification of the same model [6,7]. In the case of the x_3, the three curves overlap, because parameters corresponding to (3) are not modified, hence nominal values were used. Those parameters are neglected in accordance with the previous sensitivity analysis [16]. Compared to the fixed identified parameter set by varying key parameters of the system, the observers are able to catch additional dynamics and track the measurements more accurately. This is particularly the case where the first administration of the chemotherapeutic drug made the tumor to shrink its volume close to zero. In fact, zero volumes were measured–namely, zero volumes are represented in the data–due to such small volumes cannot be measured accurately using calipers, albeit after a certain period of time, the tumor began to grow again. This phenomenon is observed in Figures 5 and 7, where the identified sets could not model the first growth period. In Figure 9 the first growth occurs, but the discrepancy between the measurements and the model is still large. The even-numbered Figures 6–10. show how the observers tackled the problem of tracking the first growth. It is particularly noticable in the case of the MHE, which started from a high growth rate (parameter a) and reduced it gradually to be able to maintain the tumor volume close to zero. Furthermore, it can be seen that the parameter a remains close to the identified value. However, at distinct time instances, greater divergence can be experienced. These divergences usually occur when the tumor exhibits a large growth rate. It can be observed that the identified models have a better fit when the tumor didn't shrink to zero volumes after initial growth. This phenomena can be observed in Figures 11 and 13.

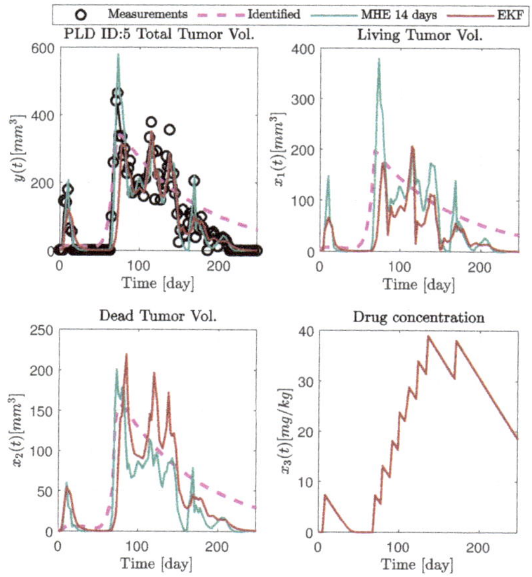

Figure 5. Estimated states of PLD5 mouse.

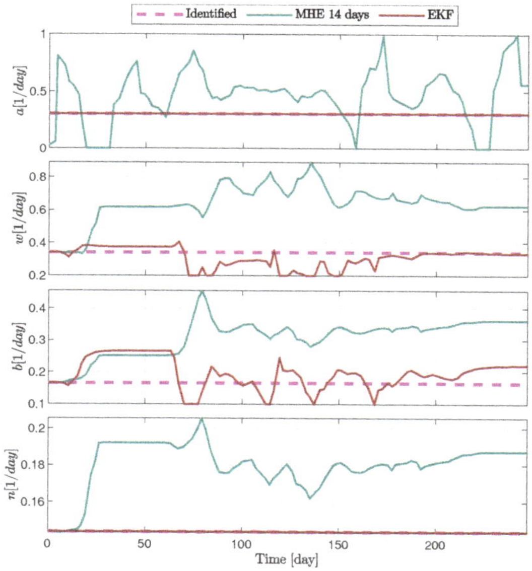

Figure 6. Estimated parameters of PLD5 mouse.

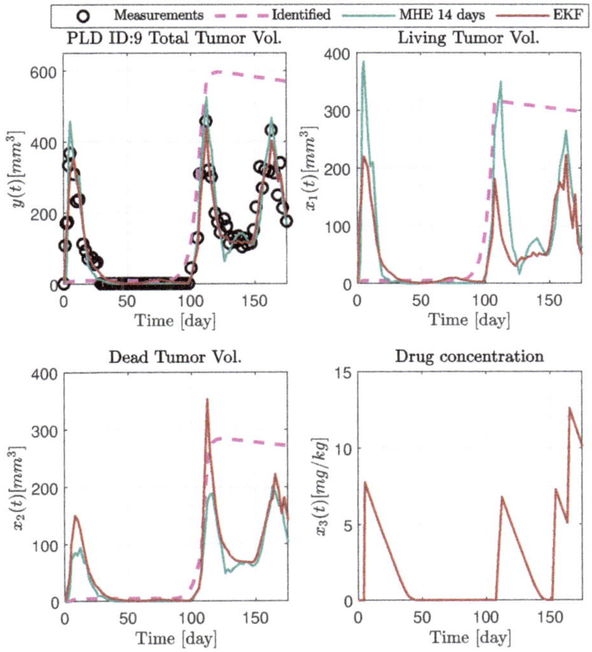

Figure 7. Estimated states of PLD9 mouse.

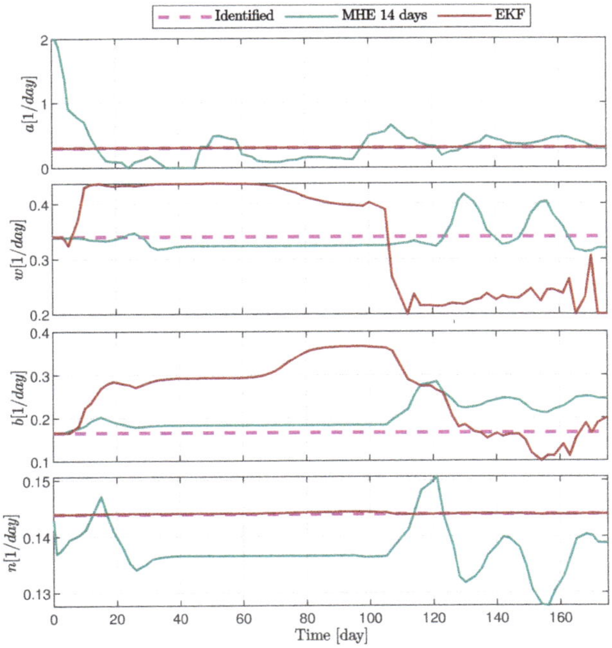

Figure 8. Estimated parameters of PLD9 mouse.

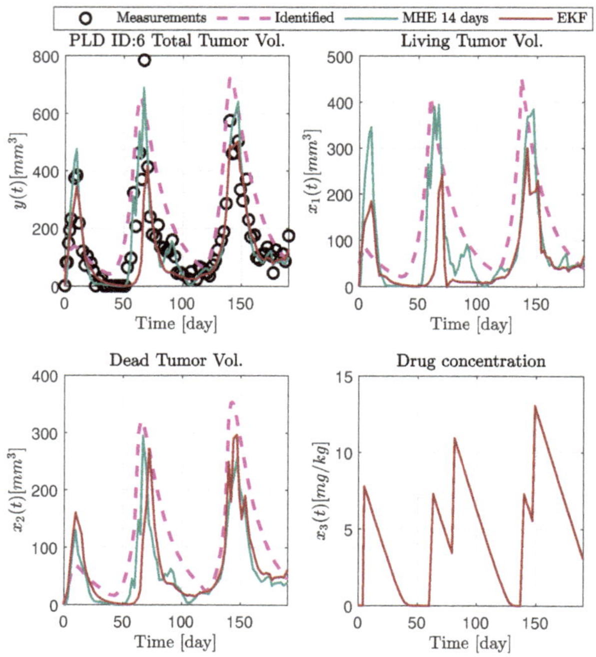

Figure 9. Estimated states of PLD6 mouse.

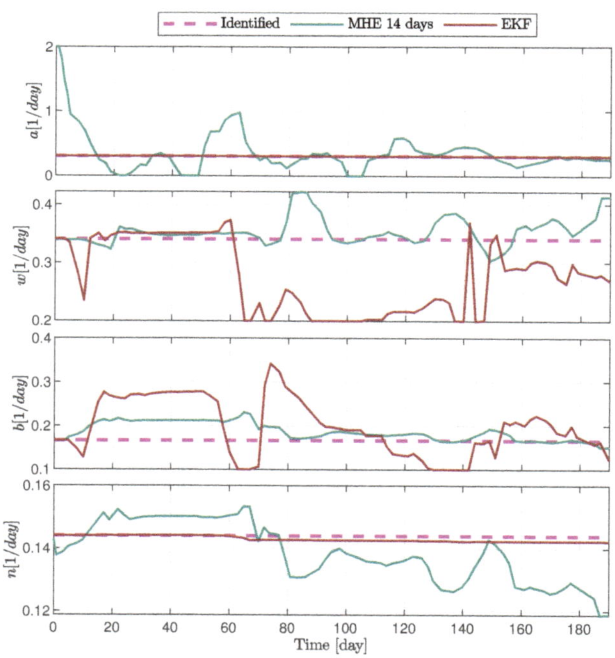

Figure 10. Estimated parameters of PLD6 mouse.

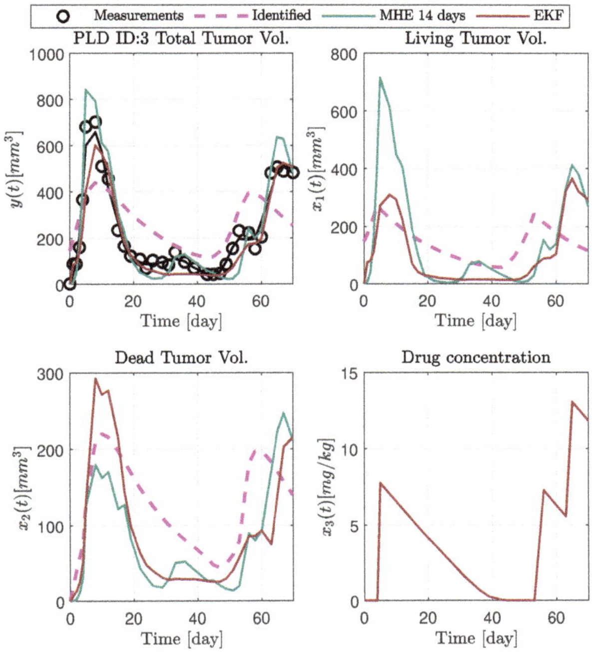

Figure 11. Estimated states of PLD3 mouse.

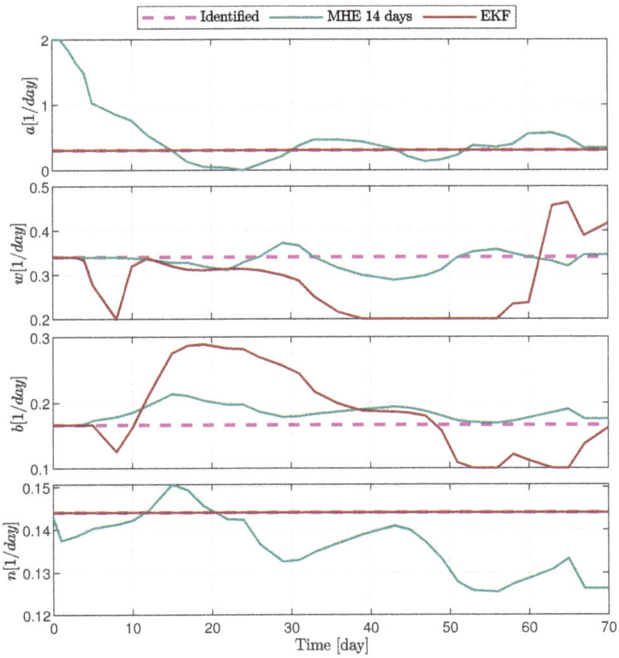

Figure 12. Estimated parameters of PLD3 mouse.

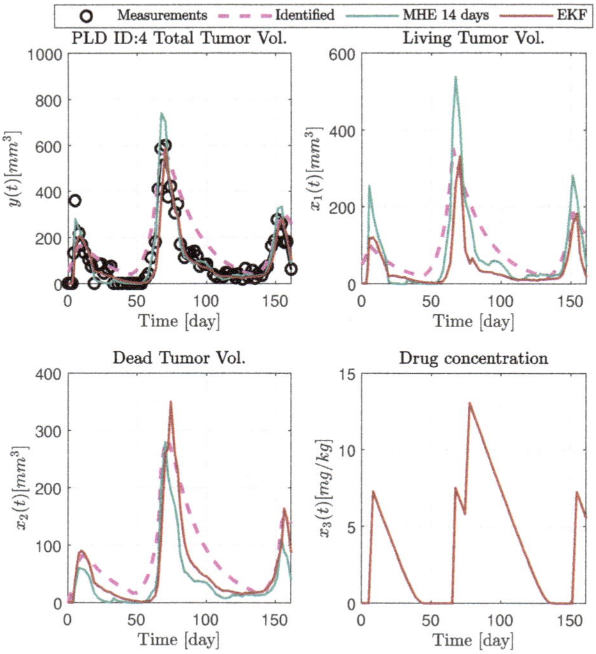

Figure 13. Estimated states of PLD4 mouse.

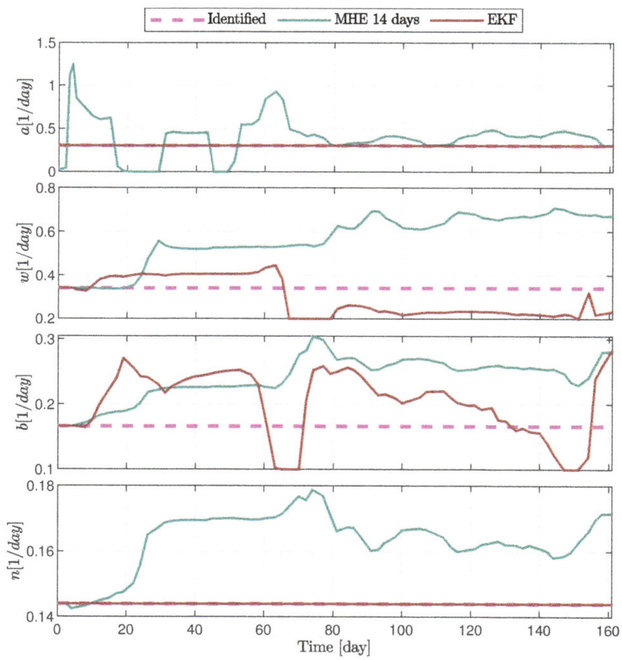

Figure 14. Estimated parameters of PLD4 mouse.

4. Discussion

The goal was to evaluate the state and parameter estimation capabilities of an MHE and an EKF which has been applied as a benchmark. In general, state and parameter estimation in the case of physiological processes have major difficulties which are presented in this study related to tumor growth estimation as well. Furthermore, the problem formulation makes it even more difficult. In the model, only the sum of the living and dead tumor volumes can be measured, often inaccurately. This limited amount of data can easily lead to observability issues mostly coming from the fact that several state and parameter combinations may result in the same output. In order to limit these issues, we applied two key components: penalizing the velocity of changing of state variables and estimated parameters, moreover, we have selected a branch of key parameters to be estimated based on sensitivity analysis. Namely, the cost function of the MHE made it possible to use additional constraints and penalize the change of parameters. The penalization was designed to be close to the physiological rate of changes to be as as realisic as possible. According to the results, we have successfully limited the set of possible state and parameter sets leading to the same output without a downturn in the estimation accuracy.

The scenario for the comparison was designed to be biased arbitrary. The results show that even with optimizing the tuning of the EKF, and not taking into account the initial state of the MHE as an additional free variable, a moving horizon estimation based observer can still outperform the former EKF. The MHE-based observers achieved lower RMSE until the horizon length was shorter than 30 days. Using horizons around 20 days, there can be cases, where the EKF outperforms the MHE, e.g., the estimation of PLD9. At 7 days the difference in RMSE is around 30 [mm^3], and the MHE has lower RMSE for each mouse. The inferior performance of the EKF can rise from the low gains for parameters a and n. Since the system is most sensitive to those parameters, having larger gains could result in divergence in the case of certain patients due to the high interpatient variability. The estimation accuracy of the MHE deteriorated in a quasilinear manner with respect to the window length. The increased window lengths resulted in an increased RMSE of the tumor volumes because the effects of the structural mismatch became dominant. In general, a longer window would mean that the given parameter set is valid for more data. However, the predictive capabilities are crucial in therapy optimization, thus the determination of the optimal choice for closed-loop applications is future work. The runtime of the observers are greatly different. The EKF averages 2.6 ms, while the MHE ranges from 300 ms to 413 ms depending on the horizon length. However, runtime is not a cornerstone as the injections and measurements are done once a day at a maximum rate. An important aspect of the study is that the observers are evaluated on real laboratory data, where two major difficulties are present: structural mismatch and intrapatient variability.

The results indicate the great importance of applying observers for therapy optimization algorithms. Without constant feedback about the behavior of the patient the model using only a single parameter set can introduce large discrepancies. As the tumor shifts phases, and as the heterogenity of the tumor changes its behavior changes as well. These shifts in physiology can be tracked with the observers–particularly well with the MHE–. These changes can be interpreted as trends in the parameters. Additionally, despite the optimized tuning of the EKF the MHE shows a smoother estimation of the parameters and a more accurate approximation of the measurements. These attributes facilitate the application of the MHE in closed-loop algorithms.

5. Conclusions

In this paper, an MHE-based observer and an EKF was designed and compared for therapy optimization and state-feedback kind of control applications. When using time horizons less than 28 days, the MHE based observer had superior performance—compared to the EKF—in tracking the measurements. Increased horizon length deteriorated the approximations as the actual parameter set had to describe longer sections, where structural mismatch and/or parameter variation occurs due to the change in the physiology of the tumor. To evaluate the performance more thoroughly advanced measurement technology is needed to distinguish live and dead tumor cells and even drug concentration. Further improvements can be done on the fine-tuning of the parameters in the cost function. Future work includes the integration of the MHE into closed-loop systems.

Author Contributions: Conceptualization, M.S., G.E., D.A.D. and L.K.; methodology, M.S., G.E., D.A.D.; writing—original draft preparation, M.S., G.E.; writing—review and editing, M.S., G.E., D.A.D., I.R., L.K.; supervision, G.E., D.A.D., L.K. All authors have read and agreed to the published version of the manuscript.

Funding: This project has received funding from the European Research Council (ERC) under the European Union's Horizon 2020 research and innovation programme (grant agreement No 679681). Project no. 2019-1.3.1-KK-2019-00007. has been implemented with the support provided from the National Research, Development and Innovation Fund of Hungary, financed under the 2019-1.3.1-KK funding scheme. M. Siket was supported by the ÚNKP-19-3 and ÚNKP-20-3 New National Excellence Program of the Ministry for Innovation and Technology.

Acknowledgments: The Authors would like to thank to András Füredi and Gergő Szakács from the Membrane Protein Research Group of the Research Centre for Natural Sciences for providing the experimental data and the many useful advices they made. The Authors would also like to thank the support of the Applied Informatics and Applied Mathematics Doctoral School.

Conflicts of Interest: The authors declare no conflict of interest.

Ethical Statement: The data of animal experiments has been provided by the Membrane Protein Research Group of the Research Centre for Natural Sciences. In this project, approval of any ethical committee has not been necessary.

Abbreviations

The following abbreviations are used in this manuscript:

EKF	Extended Kalman filter
MHE	moving horizon estimation
RMSE	root-mean-square error

References

1. WHO: Cancer. Available online: https://www.who.int/news-room/fact-sheets/detail/cancer (accessed on 11 March 2020).
2. Ren, H.P.; Yang, Y.; Baptista, M.S.; Grebogi, C. Tumour chemotherapy strategy based on impulse control theory. *Philos. Trans. Math. Phys. Eng. Sci.* **2017**, *375*. [CrossRef] [PubMed]
3. Drexler, D.A.; Ferenci, T.; Kovács, L. Extended tumor growth model for combined therapy. In Proceedings of the 2019 IEEE International Conference on Systems, Man and Cybernetics (SMC), Bari, Italy, 6–9 October 2019; pp. 886–891. [CrossRef]
4. Hahnfeldt, P.; Panigrahy, D.; Folkman, J.; Hlatky, L. Tumor Development under Angiogenic Signaling. *Cancer Res.* **1999**, *59*, 4770. [PubMed]
5. d'Onofrio, A.; Gandolfi, A. Tumour eradication by antiangiogenic therapy: Analysis and extensions of the model by Hahnfeldt et al. (1999). *Math. Biosci.* **2004**, *191*, 159–184. [CrossRef] [PubMed]
6. Drexler, D.A.; Ferenci, T.; Lovrics, A.; Kovács, L. Modeling of tumor growth incorporating the effect of pegylated liposomal doxorubicin. In Proceedings of the 2019 IEEE 23nd International Conference on Intelligent Engineering Systems, Gödöllő, Hungary, 25–27 April 2019; pp. 369–373.
7. Drexler, D.A.; Ferenci, T.; Lovrics, A.; Kovács, L. Tumor dynamics modeling based on formal reaction kinetics. *Acta Polytech. Hung.* **2019**, *16*, 31–44. [CrossRef]
8. Kuznetsov, M.; Kolobov, A. Investigation of solid tumor progression with account of proliferation/migration dichotomy via Darwinian mathematical model. *J. Math. Biol.* **2020**, *80*. [CrossRef] [PubMed]
9. Kuznetsov, M. Mathematical Modeling Shows That the Responseof a Solid Tumor to Antiangiogenic TherapyDepends on the Type of Growth. *Mathematics* **2020**, *8*, 760. [CrossRef]
10. Rokhforoz, P.; Jamshidi, A.A.; Sarvestani, N.N. Adaptive robust control of cancer chemotherapy with extended Kalman filter observer. *Inform. Med. Unlocked* **2017**, *8*, 1–7. [CrossRef]

11. Costa, J.M.; Orlande, H.R.; Velho, H.F.C.; de Pinho, S.T.; Dulikravich, G.S.; Cotta, R.M.; da Cunha Neto, S.H. Estimation of Tumor Size Evolution Using Particle Filters. *J. Comput. Biol.* **2015**, *22*, 649–665. [CrossRef] [PubMed]
12. Sápi, J.; Drexler, D.; Harmati, I.; Sápi, Z.; Kovács, L. Qualitative analysis of tumor growth model under antiangiogenic therapy—Choosing the effective operating point and design parameters for controller design. *Optim. Control Appl. Methods* **2016**, *37*, 848–866. [CrossRef]
13. Chen, T.; Kirkby, N.; Jena, R. Optimal dosing of cancer chemotherapy using model predictive control and moving horizon state/parameter estimation. *Comput. Methods Programs Biomed.* **2012**, *108*. [CrossRef] [PubMed]
14. Füredi, A.; Szebényi, K.; Tóth, S.; Cserepes, M.; Hámori, L.; Nagy, V.; Karai, E.; Vajdovich, P.; Imre, T.; Szabó, P.; et al. Pegylated liposomal formulation of doxorubicin overcomes drug resistance in a genetically engineered mouse model of breast cancer. *J. Control Release* **2017**, *261*, 287–296. [CrossRef] [PubMed]
15. Drexler, D.A.; Nagy, I.; Romanovski, V.; Tóth, J.; Kovács, L. Qualitative analysis of a closed-loop model of tumor growth control. In Proceedings of the 18th IEEE International Symposium on Computational Intelligence and Informatics, Budapest, Hungary, 21–22 November 2018; pp. 329–334.
16. Siket, M.; Eigner, G.; Kovács, L. Sensitivity and identifiability analysis of a third-order tumor growth model. In Proceedings of the IEEE 15th International Conference on System of Systems Engineering, Budapest, Hungary, 2–4 June 2020.
17. Villaverde, A.F.; Barreiro, A. Identifiability of Large Nonlinear Biochemical Networks. *MATCH Commun. Math. Comput. Chem.* **2016**, *96*, 259–296.
18. Iman, R.L. Latin Hypercube Sampling. In *Wiley StatsRef: Statistics Reference Online*; American Cancer Society: Atlanta, GA, USA, 2014. [CrossRef]

Publisher's Note: MDPI stays neutral with regard to jurisdictional claims in published maps and institutional affiliations.

applied sciences

MDPI

Article

A Review of Sample and Hold Systems and Design of a New Fractional Algorithm

Manuel Duarte Ortigueira [1] and José Tenreiro Machado [2,*]

1 CTS-UNINOVA and DEE of NOVA School of Science and Technology, Quinta da Torre, 2829-516 Caparica, Portugal; mdo@fct.unl.pt

2 Department of Electrical Engineering, Institute of Engineering, Polytechnic of Porto, 4249-015 Porto, Portugal

* Correspondence: jtm@isep.ipp.pt

Received: 3 October 2020; Accepted: 18 October 2020; Published: 21 October 2020

Abstract: Digital systems require sample and hold (S&H) systems to perform the conversion from analog to digital and vice versa. Besides the standard zero and first order holds, we find in the literature other versions, namely the fractional and exponential order holds, involving parameters that can be tuned to produce a superior performance. This paper reviews the fundamental concepts associated with the S&H and proposes a new fractional version. The systems are modeled both in the time and Laplace domains. The new S&H stemming from fractional calculus generalizes these devices. The different S&H systems are compared in the frequency domain and their relationships visualized by means of hierarchical clustering and multidimensional scaling representations. The novel strategy allows a better understanding of the possibilities and limitations of S&H systems.

Keywords: sample and hold; fractional calculus; hierarchical clustering; multidimensional scaling; distances

1. Introduction

Fractional calculus (FC) generalizes the classical theory of differential calculus by including the concept of integrals and derivatives of real or complex order [1,2]. Many systems in physics and engineering can be generalized in the light of the FC concepts. However, a number of applications still remain to be fully explored under the light of FC. The areas of signal processing, dynamical systems and control has received considerable attention and the results show that the adoption of FC is clearly a fruitful research strategy [3–6].

Sample and hold (S&H) systems are commonly applied in mix analog/digital systems, being the standard procedure in discrete time systems to adopt zero-order holds (S&H-0) [7–10]. The devices, are simple to construct and 'hold' the signal sample during one time interval. They are a basic tool of the analog to digital (A/D) conversion. In the reverse operation (D/A) other types of S&H are used. For instance, the first-order hold (S&H-1) reconstructs signals as a piecewise linear approximations and may yield better fit at the price of a higher complexity. Other S&H are the fractional- and exponential- order holds, abbreviated as S&H-β and S&H-exp, with parameters $0 \leq \beta \leq 1$ and $\tau > 0$, respectively, that also improve the technique at the cost of a slight additional complexity [11–14]. In fact, we can take advantage of the extra parameter β and τ by tuning their values for a particular application. Nonetheless, we observe the introduction of high frequency artifacts, that may pose difficulties in some applications. We can reduce this kind of distortion by using a modification of some of the above S&H systems. This change consists in introducing a delay of one sampling interval until a new sample appears. Therefore, instead of an extrapolation, as above, we are doing an interpolation [10,15]. Such procedures are very important in control [16,17] and telecommunications [9,18]. These topics are often neglected in practical applications. On one hand, the difference between theoretical and engineering implementations, due to physical restrictions, are overlooked. On the other hand, the possibilities for further improving the S&H systems, particularly when considering FC, remain to be explored.

This paper starts by reviewing the fundamental concepts associated with the S&H systems and proposes one generalization in the perspective of FC. The classical and fractional S&H are then compared based on their frequency responses. For that purpose we adopt the Euclidean and Canberra distances [19]. The distinct realizations of S&H are then represented using the Hierarchical Clustering (HC) and multidimensional scaling (MDS) computational tools [20–23]. Both schemes rely on constructing graphical representations of the items under study so that users can visualized their relationships. Items are 'leafs' of structures such as dendrogram or trees, for the HC, and as points in some abstract space, for the MDS. In both cases similar items are represented closely and vice-versa.

Bearing these ideas in mind the manuscript is organized as follows. Section 2 introduces the fundamental concepts concerning S&H. Section 3 discusses real-world implementations of S&H systems and proposes a fractional-order generalization. Section 4 analyses the transfer functions of the S&H systems under discussion. Finally, Section 5 outlines the conclusions.

2. Fundamental Concepts

With present day digital computer systems, and in areas such as signal processing and control, the conversion of an analog signal into a digital version requires sampling and quantisation. This action is called the A/D conversion (analog to digital conversion) and its inverse is the D/A conversion. In general, there are infinite continuous signals that can generate a given discrete signal. However, when the signals are bandlimited, meaning that they are of bounded support in the frequency domain, we can recover the original signal.

Definition 1. *The process of associating a discrete signal to a continuous one such that*

1. *the domain of the discrete signal is a discrete subset of the domain of the continuous signal,*
2. *the range of the discrete signal is also a discrete subset of the range of the continuous signal,*
3. *both signals, discrete and continuous, assume the same values at the intersection domains,*

is named sampling [6].

The inverse operation, that is, the calculation of intermediate continuous values based on the discrete samples, is called interpolation or reconstruction. Ideally, the sampling operation is performed by the delta comb that we introduce in the follow-up.

Definition 2. *Let F and $T = \frac{1}{F}$ be two positive real constants. We define an almost linear time scale (ALTS) by [24]*

$$t_n = n \cdot T + \tau_n, \quad T = 1/F, \quad n \in \mathbb{Z}, \tag{1}$$

where τ_n a zero mean sequence such that

$$0 \leq |\tau_n| < \frac{T}{4}. \tag{2}$$

The constant F is called sampling frequency , or density of the time scale, while its inverse T is the average graininess or sampling interval [25].

Remark 1. *In some applications, we prefer a half sampling interval delayed sequence $t_n = (n + 1/2) \cdot T + \tau_n$.*

Definition 3. *On the above ALTS (1) we define a uniform (in amplitude) almost periodic delta comb by*

$$p(t) = \sum_{n=-\infty}^{+\infty} \delta(t - t_n), \qquad t \in \mathbb{R}, \tag{3}$$

that is considered as the ideal sampler, since its multiplication by any signal, $y(t)$, gives its sampled version

$$y_p(t) = y(t)p(t) = \sum_{n=-\infty}^{+\infty} y(t_n)\delta(t - t_n), \qquad t \in \mathbb{R}. \tag{4}$$

This distribution has the Fourier transform (FT)

$$Y_p(\omega) = \sum_{n=-\infty}^{+\infty} y(t_n)e^{-j\omega t_n}, \qquad \omega \in \mathbb{R}, \tag{5}$$

where $j = \sqrt{-1}$ and

$$Y_p(\omega) = \int_{-\infty}^{\infty} y_p(t)e^{-j\omega t}\mathrm{d}t.$$

Theorem 1. *Let $\Omega \in \mathbb{R}^+$ be the density of an odd ALTS, that is, so that $\omega_n = -\omega_{-n}$, given by $\omega_n = n\Omega + \nu_n$, with $|\nu_n| < \Omega/4$, $n \in \mathbb{Z}$. The FT of the comb (3) in the time domain is another comb [24] in the frequency domain given by:*

$$C(\omega) = 2\pi \sum_{n=-\infty}^{+\infty} C_n \delta\left(\omega - \omega_n\right), \tag{6}$$

where the coefficients,

$$C_k = \frac{1}{T} \lim_{N\to\infty} \frac{1}{2N+1} \sum_{n=-N}^{N} \mathrm{e}^{-j\omega_k t_n},$$

form a bounded sequence and

$$\frac{\Omega T}{2\pi} = 1. \tag{7}$$

This has as consequence that

$$Y_p(\omega) = \sum_{n=-\infty}^{+\infty} C_n Y(\omega - \omega_n), \tag{8}$$

showing that $Y_p(\omega)$ is an almost periodic function that results from the almost periodic repetition of $Y(\omega)$. Therefore, if $y(t)$ is not bandlimited, or if Ω is not high enough, then two consecutive repeated images of $Y(\omega)$ may overlap (phenomenon called aliasing) and it is not possible to recover the original signal by low-pass filtering. Unless something else is specified, we will assume that our signals are bandlimited in the follow-up. If the bandwidth is W, then we conclude that $2W < \Omega - \Omega/4 < \omega_1$. Otherwise, if we allow a spectral superposition, $2W > \omega_1$, then we cannot recover the original signal by low-pass filtering.

In the uniform sampling case $t_n = nT$ and $\omega_n = n\Omega$, and we must have $2W < \Omega$, stating the well known Nyquist rule: the sampling interval must be less than one half on the inverse of the higher frequency of the signal. The relation (7) shows that reducing the average distance between the impulses in the time domain (4) corresponds to increasing the distances between the replicas in the frequency domain (8).

Theorem 2. *Let $x(t)$ be a continuous-time bandlimited signal with Fourier transform, $X(\omega)$, that is null outside the band $|\omega| < W \in \mathbb{R}^+$. We can recover $x(t)$ from its samples $x(t_n)$ with a low-pass filtering provided that the filter frequency response is constant in the band of the signal. In the case of an ideal low-pass filter with bandwidth W, the output reads [24]*

$$x(t) = \sum_{n=-\infty}^{\infty} x\left(t_n\right) \frac{\sin\left[W\left(t-t_n\right)\right]}{(\Omega/2)\left(t-t_n\right)}. \tag{9}$$

This theorem constitutes a generalization of the classical sampling theorem. This results deserves an interpretation in terms of linear systems. Let us consider that

$$g(t) = \frac{\sin(Wt)}{(\Omega/2)t}, \quad t \in \mathbb{R}, \tag{10}$$

is the impulse response of a low-pass linear system and that the input to such system is

$$x_p(t) = \sum_{n=-\infty}^{+\infty} x(t_n)\delta(t-t_n).$$

Then the output will be given by the expression in (9). The signal $g(t)$ is called the interpolating function and its FT is a rectangle which gives a justification to its name of 'ideal low-pass filter'.

If the t_n sequence is uniform, that is, if $\tau_n \equiv 0$, then

$$x(t) = \sum_{n=-\infty}^{\infty} x(nT) \frac{\sin\left[WT\left(\frac{t}{T}-n\right)\right]}{\pi\left(\frac{t}{T}-n\right)}, \tag{11}$$

that states the usual sampling theorem, where $WT < \pi$.

3. Sampling and Reconstruction

We reviewed the ideal sampling having as base the comb. In practice, we cannot implement the ideal Dirac impulses, but we can approximate them by narrow high rectangular pulses, $r(t)$, with area equal to 1. Nonetheless, we are interested in having the values $x(t_n)$, rather than $x(t_n)\delta(t-t_n)$ or its realistic version $x(t_n)r(t-t_n)$. Therefore, we need to adopt a system that keeps, i.e. 'memorizes', the value $x(t_n)$ during a sufficient time for the computer system to be able to read the value.

The standard sampling is performed by a zero order hold. This scheme transforms the signal $x(t)$ in a sequence of rectangles with height equal to $x(t_n)$ and width less or equal to the sampling interval. This procedure, called pulse amplitude modulation (usually abbreviated as PAM) in telecommunications [18], introduces an amplitude distortion and increases of the bandwidth that are corrected by a lowpass equalizer filter. However, there are several distinct ways of implementing the reconstruction and we review a few in the sequel.

3.1. The S&H Reconstruction by Extrapolation

Equation (9), is important from the mathematical point of view, but it is not of practical usefulness, since it requires the knowledge of all the signal values $x(t_n)$, both at the past and future. This is a consequence of the fact that $g(t)$ is not the impulse response of a causal system. To avoid such difficulty we use causal approximations generically called S&H systems. The most known are the zero- and one-order, but we can mention also the exponential hold. Let us analyze several approaches.

3.1.1. The Zero Order Sample and Hold

The output signal of a zero-order hold (S&H-0) has a zero slope between two consecutive sampling instants. Therefore, the S&H-0 circuit retains the measured value for a sampling instant at that level until the next sampling instant. The impulse response is a rectangle:

$$\hat{x}(t) = x(t_n)\left[u(t-t_n) - u(t-t_{n+1})\right],$$

for $t_n \leq t < t_{n+1}$, where $u(t)$ is the Heaviside unit step.

3.1.2. The First-Order Sample and Hold

The first-order hold (S&H-1) exhibits an impulse response that has a constant slope between two consecutive sampling instants. This slope is the incremental ratio determined by the values of the two preceding samples. Thus, the first-order-hold predicts the response over the sampling interval from t_n to t_{n+1} as a ramp with the slope determined by the signal values at the two consecutive time instants. The impulse response is

$$\hat{x}(t) = x(t_n)\left[u(t-t_n) - u(t-t_{n+1})\right] + (t-t_n)\frac{x(t_n) - x(t_{n-1})}{(t_n - t_{n-1})}\left[u(t-t_n) - u(t-t_{n+1})\right],$$

for $t_n \leq t < t_{n+1}$.

3.1.3. The Fractional-Order Sample and Hold

If we calculate the the arithmetic average between the S&H-0 and S&H-1, with weight factors $1-\beta$ and β, $0 \leq \beta \leq 1$, respectively, then we obtain the fractional-hold (S&H-β) [12–14].

The impulse response of the S&H-β is:

$$\hat{x}(t) = x(t_n)\left[u(t-t_n) - u(t-t_{n+1})\right] + \beta(t-t_n)\frac{x(t_n) - x(t_{n-1})}{(t_n - t_{n-1})}\left[u(t-t_n) - u(t-t_{n+1})\right],$$

for $t_n \leq t < t_{n+1}$. Therefore, we obtain the S&H-0 and S&H-1 for $\beta = 0$ and $\beta = 1$, respectively.

3.1.4. The exponential order sample and hold

The exponential order hold S&H-exp is interesting because it has a simple electrical implementation based on an RC low-pass circuit [11]. The impulse response of the S&H-exp is:

$$\hat{x}(t) = x(t_n)\left[u(t-t_n) - u(t-t_{n+1})\right] + \left[1 - e^{-\frac{(t-t_n)}{\tau}}\right]\frac{x(t_n) - x(t_{n-1})}{(t_n - t_{n-1})}\left[u(t-t_n) - u(t-t_{n+1})\right],$$

for $t_n \leq t < t_{n+1}$, where τ is the time constant. This expression is slightly different from the one presented in [11]. However, the previous expression is preferable because gives the S&H-0 when $\tau \to \infty$.

3.1.5. The α-Order Sample and Hold

Given the expressions for the S&H-0 and S&H-1 we can propose a generalization as follows.

Definition 4. *The α-order (S&H-α) is formulated as:*

$$\hat{x}(t) = x(t_n) + (t-t_n)^{\alpha}\frac{x(t_n) - x(t_{n-1})}{(t_n - t_{n-1})^{\alpha}}, \tag{12}$$

for $t_n \leq t < t_{n+1}$, *where* $\alpha \in \mathbb{R}^+$.

The impulse response is given by:

$$\hat{x}(t) = x(t_n)\left[u(t-t_n) - u(t-t_{n+1})\right] + (t-t_n)^{\alpha}\frac{x(t_n)-x(t_{n-1})}{(t_n-t_{n-1})^{\alpha}}\left[u(t-t_n) - u(t-t_{n+1})\right], \quad (13)$$

for $t_n \leq t < t_{n+1}$.

3.2. The S&H Reconstruction by Interpolation

The S&H tools presented in the previous sub-section have a drawback, since they may lead to a discontinuous signal. The presence of discontinuities transforms the original bandlimited signal into fullband as we can observe in Figure 1. We can avoid this difficulty by implementing some kind of interpolation instead of having an extrapolation. However, this strategy has a price, since it involves the problem of a one sample delay. In many applications this delay has no visible effect. For example, in speech processing, with a sampling interval equal to 125 μs, the effect of such delay is negligible. Nevertheless, not all the S&H techniques discussed above can be used. In fact, the S&H-0 leads always to discontinuities. The exponential hold has also that problem, unless we modify the parameter τ from sample to sample, which is not easily implementable. This problem does not happen with the proposed general S&H-α, where we only have to make a slight modification.

Definition 5. *The* α*-order (S&H-*α*) with delay is expressed as:*

$$\hat{x}(t) = x(t_n) + (t-t_n)^{\alpha}\frac{x(t_{n+1})-x(t_n)}{(t_{n+1}-t_n)^{\alpha}}, \quad t_n \leq t \leq t_{n+1}. \quad (14)$$

The impulse response is given by:

$$\hat{x}(t) = x(t_n)\left[u(t-t_n) - u(t-t_{n+1})\right] + (t-t_n)^{\alpha}\frac{x(t_{n+1})-x(t_n)}{(t_{n+1}-t_n)^{\alpha}}\left[u(t-t_n) - u(t-t_{n+1})\right]. \quad (15)$$

The S&H-α with delay (14) leads to a continuous signal, having as consequence that the amplitude spectrum decreases, at least, as fast as ω^{-2} when $\omega \to \infty$.

Example 1. *On the left of Figure 2 we depict the extrapolations obtained with the S&H-0, S&H-*α *(with* $\alpha = 0.3$ *and* $\alpha = 0.7$*), S&H-1 and S&H-exp (with* $\tau = 200$*). On the right, we show the interpolations resulting from the S&H-*α *(with* $\alpha = 0.3$ *and* $\alpha = 0.7$*) and the S&H-1.*

The difference between the two kinds of procedures is clear, as well as their distinct behavior.

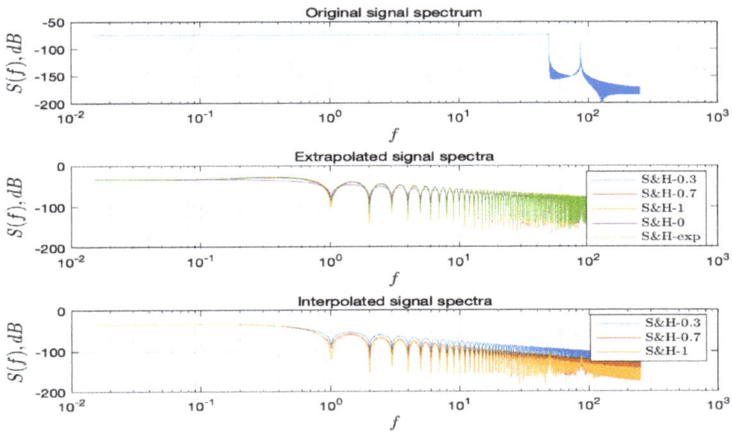

Figure 1. Frequency response of the extrapolations of the S&H-0, S&H-α (with $\alpha = 0.3$ and $\alpha = 0.7$), S&H-1 and S&H-exp (with $\tau = 200$) and interpolations with S&H-α (with $\alpha = 0.3$ and $\alpha = 0.7$) and the S&H-1.

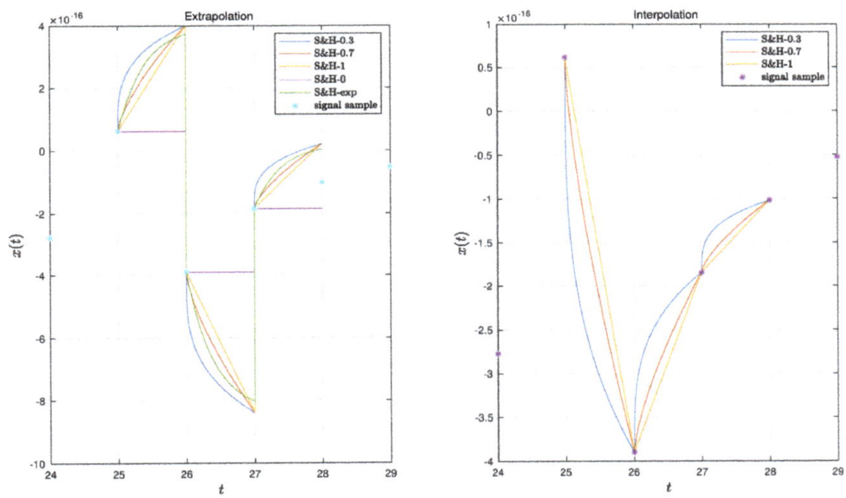

Figure 2. Time response of the extrapolations of the S&H-0 , S&H-α (with $\alpha = 0.3$ and $\alpha = 0.7$), S&H-1 and S&H-exp (with $\tau = 200$) and interpolations with S&H-α (with $\alpha = 0.3$ and $\alpha = 0.7$) and the S&H-1.

4. The Transfer Functions of the S&H

In the previous section we introduced several S&H and presented the corresponding impulse responses. These have a general form

$$\hat{x}(t) = ar(t) + bg(t), \tag{16}$$

where a and b are real constants, $r(t) = u(t - t_n) - u(t - t_{n+1})$ and $g(t)$ is a short duration signal (i.e., a pulse). The first term in (16) is a rectangle and its Laplace transform is

$$\mathcal{L}\left[r(t)\right] = \frac{e^{-st_n} - e^{-st_{n+1}}}{s}, \qquad s \in \mathbb{C}, \tag{17}$$

where $\mathcal{L}$ and s denote the Laplace operator and variable, respectively.

In the case of a uniform time scale, $t_n = nT$,

$$r(t) = u(t - nT) - u(t - (n+1)T),$$

Equation (17) gives

$$\mathcal{L}\left[r(t)\right] = \frac{e^{-snT} - e^{-s(n+1)T}}{s}, \qquad s \in \mathbb{C}. \tag{18}$$

The second term in (16) depends on the S&H method. For the fractional case (that includes $\alpha = 1$), that is, for relations, (13) and (15), we obtain

$$\mathcal{L}\left[(t - t_n)^\alpha r(t)\right] = e^{-st_n} \int_0^{t_{n+1}-t_n} t^\alpha e^{-st} dt, \qquad s \in \mathbb{C}. \tag{19}$$

Let us introduce the lower incomplete gamma function defined as:

$$\gamma(\alpha, v) = \int_0^v \tau^{\alpha-1} e^{-\tau} \mathrm{d}\tau, \quad \alpha, v > 0,$$

that verifies the relation

$$\gamma(\alpha + 1, v) = \alpha\gamma(\alpha, v) - v^\alpha e^{-v}.$$

We can write

$$\int_0^{t_{n+1}-t_n} t^\alpha e^{-st} dt = \alpha \frac{\gamma(\alpha, (t_{n+1} - t_n)s)}{s^{\alpha+1}} - \frac{t_{n+1} - t_n}{s} e^{-(t_{n+1}-t_n)s}, \quad Re(s) > 0,$$

that leads to

$$\mathcal{L}\left[(t - t_n)^\alpha r(t)\right] = \alpha \frac{\gamma(\alpha, (t_{n+1} - t_n)s)}{s^{\alpha+1}} e^{-st_n} - \frac{t_{n+1} - t_n}{s} e^{-st_{n+1}}, \quad Re(s) > 0. \tag{20}$$

For a uniform time scale Equation (20) yields

$$\mathcal{L}\left[(t - nT)^\alpha r(t)\right] = \alpha \frac{\gamma(\alpha, Ts)}{s^{\alpha+1}} e^{-snT} - \frac{T}{s} e^{-s(n+1)T}, \quad Re(s) > 0. \tag{21}$$

In the particular case of $\alpha = 1$, we obtain $\gamma(1, v) = \int_0^v e^{-\tau} \mathrm{d}\tau = 1 - e^{-v}$, resulting

$$\mathcal{L}\left[(t - t_n) r(t)\right] = \frac{e^{-st_n} - e^{-st_{n+1}}}{s^2} - \frac{t_{n+1} - t_n}{s} e^{-st_{n+1}}, \quad Re(s) > 0, \tag{22}$$

and, therefore, to

$$\mathcal{L}\left[(t - nT) r(t)\right] = e^{-snT} \left[\frac{1 - e^{-sT}}{s^2} - \frac{T}{s} e^{-sT}\right], \quad Re(s) > 0. \tag{23}$$

Remark 2. *The choice of*

$$g(t) = \mathcal{L}^{-1}\left[\frac{e^{-st_n} - e^{-st_{n+1}}}{s^{\alpha}}\right]$$

is not of relevance, because it is not a function with finite support.

Let us recall the example discussed in the previous section. Figure 2 depicts the responses in the frequency domain.

We can compare the distinct items by means of the HC and MDS schemes. These clustering techniques allow the comparison and visualization of high dimensional data [26–29], such as in the case of controlling nonlinear systems [30].

In our case, the i-th item corresponds to the response in the frequency domain, $G_i(f)$, that is represented by a numerical array of m points. A set of N items is then compared by means of suitable distances [31]. We adopt the Euclidean and Canberra distances given by:

$$d_{ij}^{E} = \sqrt{\sum_{n=1}^{m} \left(G_i(f) - G_j(f)\right)^2}, \tag{24a}$$

$$d_{ij}^{C} = \sum_{n=1}^{m} \frac{|G_i(f) - G_j(f)|}{|G_i(f)| + |G_j(f)|}, \tag{24b}$$

where $i, j = 1, \ldots, N$, so that N denotes the total number of items under comparison.

The first distance, d^E, follows the standard logic of Euclidean spaces, while the second, d^C, distinguishes more clearly values close to zero and in general is more robust to data sets including large and small values. The MDS computational recursive algorithm requires the input of a matrix of all item-to-item distances (i.e., a N dimensional square and symmetric matrix with main diagonal of zeros) and tries to reproduce approximately that matrix by mean a plot in a $m_{MDS} < m$ dimensional space where items consist of points. Therefore, similar items lead to close points and, inversely, very different items produce points very far apart. Usually, the dimensions $m_{MDS} = 2$ or $m_{MDS} = 3$ are used, because they allow a straightforward visualization [32,33].

In our case, the $N = 9$ items consist of the frequency responses $G(f)$ described numerically in $n_f = 10^4$ sample frequencies. The frequency responses characterize the original signal as well as the S&H-0, S&H-α (with $\alpha = 0.3$ and $\alpha = 0.7$), S&H-1 and S&H-exp (with $\tau = 200$) and interpolations with S&H-α (with $\alpha = 0.3$ and $\alpha = 0.7$) and the S&H-1.

Figures 3 and 4 show the HC tree and 3-dimensional MDS plots, respectively, when we compare the the frequency response of the extrapolations of the S&H-0, S&H-α (with $\alpha = 0.3$ and $\alpha = 0.7$), S&H-1 and S&H-exp (with $\tau = 200$) and interpolations with S&H-α (with $\alpha = 0.3$ and $\alpha = 0.7$) and the S&H-1, using the Euclidean and Canberra distances.

We verify that extrapolations and interpolations (marked in red and blue, respectively) produce two distinct clusters. The extrapolation cluster is more concentrated than the interpolation, showing the behavioral in practical applications is almost independent of the adopted method, because they introduce similar discontinuities. However, in the interpolation case we see clearly an improvement when varying the parameter α.

This paper has not addressed the electronic implementation of such devices (interested readers can refer to [34–40] and references therein). Therefore, an open issue is the design of hardware systems capable of S&H advanced systems. Studies on the use of S&H-β in control systems report relevant results for improving the overall performance [41–46]. In this perspective, the construction of low cost general S&H

may be a valuable option for obtaining equipment more friendly to the environment while not posing significant financial requirements.

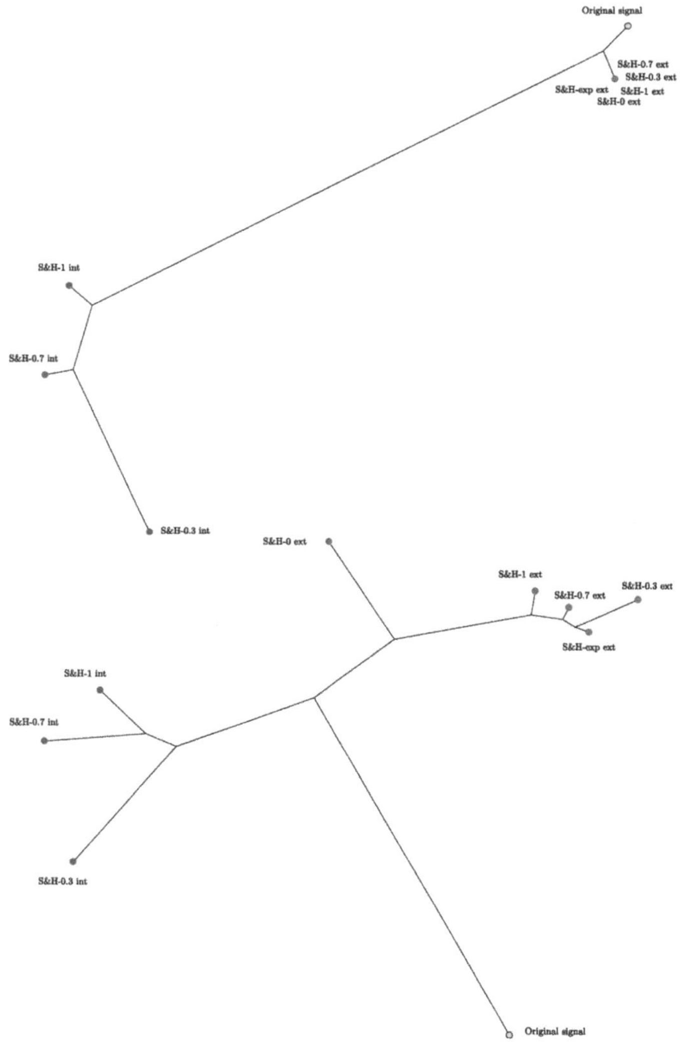

Figure 3. The HC trees using the Euclidean and Canberra distances, for the frequency response of the extrapolations of the S&H-0, S&H-α (with $\alpha = 0.3$ and $\alpha = 0.7$), S&H-1 and S&H-exp (with $\tau = 200$) and interpolations with S&H-α (with $\alpha = 0.3$ and $\alpha = 0.7$) and the S&H-1.

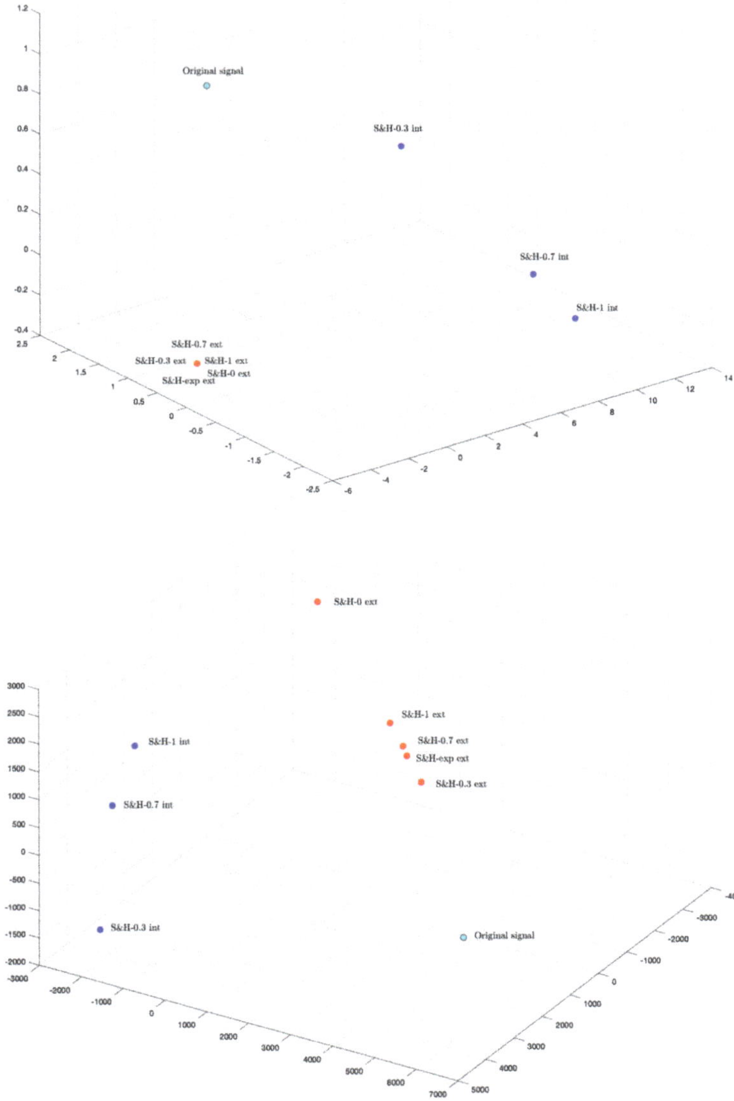

Figure 4. The MDS 3-dimensional plots using the Euclidean and Canberra distances, for the frequency response of the extrapolations of the S&H-0, S&H-α (with $\alpha = 0.3$ and $\alpha = 0.7$), S&H-1 and S&H-exp (with $\tau = 200$) and interpolations with S&H-α (with $\alpha = 0.3$ and $\alpha = 0.7$) and the S&H-1.

5. Conclusions

S&H systems are a fundamental part of present-day digital systems. In spite of their widespread use, often their characteristic are neglected and, therefore, the possibility of improving the overall performance by using advanced S&H devices is not taken into account when designing signal discrete-time processing or control systems. This paper reviewed the main mathematical ideas associated with S&H systems. The main types of S&H systems were analyzed both in the time and Laplace domains. A generalization of S&H in the viewpoint of FC was formulated so that the additional degree of freedom provided by the fractional order can be tuned for each particular application. This strategy allows designers to take advantage of a more advanced system for improving the performance of systems posing considerable dynamical challenges and requiring that all sub-system are fully optimized. Furthermore, the HC and MDS clustering techniques were used to better verify the S&H characteristics. These tools allow a novel viewpoint for the quantitative comparison between multidimensional data sets. In particular, they can be applied also for improving the real-time performance of complex control systems.

Author Contributions: These two authors contribute equally to this paper. All authors have read and agreed to the published version of the manuscript.

Funding: This work was partially funded by National Funds through the Foundation for Science and Technology of Portugal, under the projects UIDB/00066/2020.

Conflicts of Interest: The authors declare no conflict of interest.

Abbreviations

The following abbreviations are used in this manuscript:

BLT	bilateral Laplace transform
FT	Fourier transform
HC	Hierarchical clustering
LT	Laplace transform
MDS	Multidimensional scaling
S&H	Sample and hold
S&H-0	Zero order sample and hold
S&H-1	First-order sample and hold
S&H-β	Fractional-order sample and hold
S&H-exp	Exponential order sample and hold
S&H-α	α-order sample and hold

References

1. Ortigueira, M.D. *Fractional Calculus for Scientists and Engineers*; Lecture Notes in Electrical Engineering; Springer: Berlin/Heidelberg, Germany, 2011. [CrossRef]
2. Machado, J.A.T.; Mainardi, F.; Kiryakova, V. Fractional calculus: Quo vadimus? (where are we going?) *Fract. Calc. Appl. Anal.* **2015**, *18*, 495–526. [CrossRef]
3. Ortigueira, M.D. An Introduction to the Fractional Continuous-Time Linear Systems: The 21(st) Century Systems. *IEEE Circuits Syst. Mag.* **2008**, *8*, 19–26. [CrossRef]
4. Ortigueira, M.D.; Machado, J.T. Fractional derivatives: The perspective of system theory. *Mathematics* **2019**, *7*, 150. [CrossRef]
5. Valério, D.; da Costa, J. S. *An Introduction to Fractional Control*; Control Engineering; IET: London, UK, 2012.
6. Ortigueira, M.D.; Valério, D. *Fractional Signals and Systems*; De Gruyter: Berlin, Germany; Boston, MA, USA, 2020. [CrossRef]

7. Oppenheim, A.V.; Willsky, A.S.; Hamid, S. *Signals and Systems*, 2nd ed.; Prentice-Hall: Upper Saddle River, NJ, USA, 1997.
8. Garrido, A.J.; De la Sen, M.; Bárcena, R. Approximate models to describe real sampling and hold processes based on multirate sampling techniques. In Proceedings of the American Control Conference, Chicago, IL, USA, 28–30 June 2000; Volume 1, pp. 195–199. [CrossRef]
9. Kester, W. The Data Conversion Handbook. 2005. Available online: http://www.analog.com/library/analogdialogue/archives/39-06/data_conversion_handbook.html (accessed on 30 September 2020).
10. Proakis, J.G.; Manolakis, D.G. *Digital Signal Processing: Principles, Algorithms, and Applications*; Prentice Hall: Upper Saddle River, NJ, USA, 2007.
11. Pichler, H.; Pavuza, F.G.; Sommer, T. The exponential-hold circuit, properties and applications. In Proceedings of the 1993 IEEE Instrumentation and Measurement Technology Conference, Irvine, CA, USA, 18–20 May 1993; pp. 745–750. [CrossRef]
12. Sheen, I.E.; Tsai, J.S.H.; Shieh, L.S. Optimal digital redesign of continuous-time systems using fractional-order hold. *Optim. Control Appl. Meth.* **1997**, *18*, 399–422. [CrossRef]
13. Barcena, R.; de la Sen, M.; Sagastabeitia, I. Improving the stability properties of the zeros of sampled systems with fractional order hold. *IEE Proc. Control Theory Appl.* **2000**, 147, 456–464.:20000550. [CrossRef]
14. Machado, J.A.T. Fractional order modelling of fractional-order holds. *Nonlinear Dyn.* **2012**, *70*, 789–796. [CrossRef]
15. Bernhardsson, B. The predictive first order hold circuit. In Proceedings of the 29th IEEE Conference on Decision and Control, Honolulu, HI, USA, 5–7 December 1990; IEEE: Piscataway, NJ, USA, 1990.
16. de Carvalho, J.L.M. *Dynamical Systems and Automatic Control*; Prentice Hall: Hertfordshire, UK, 1993.
17. Dorf, R.C.; Bishop, R.H. *Modern Control Systems*; Pearson: London, UK, 2011.
18. Haykin, S. *Communication Systems*; John Wiley & Sons: Hoboken, NJ, USA, 2008.
19. Cha, S.-H. Comprehensive Survey on Distance/Similarity Measures between Probability Density Functions. *Int. J. Math. Model. Methods Appl. Sci.* **2007**, *1*, 300–307.
20. Torgerson, W.S. *Theory and Methods of Scaling*; Wiley: New York, NY, USA, 1958.
21. Sammo, J. A nonlinear mapping for data structure analysis. *IEEE Trans. Comput.* **1969**, *18*, 401–409. [CrossRef]
22. Kruskal, J. *Multidimensional Scaling*; Sage Publications: Newbury Park, CA, USA, 1978.
23. Kaufman, L.; Rousseeuw, P.J. *Clustering Algorithms*; Wiley-Interscience: Hoboken, NJ, USA, 2005.
24. Ortigueira, M.D. The comb signal and its Fourier transform. *Signal Process.* **2001**, *81*, 581–592. [CrossRef]
25. Ortigueira, M.D.; Torres, D.F.M.; Trujillo, J.J. Exponentials and Laplace transforms on nonuniform time scales. *Commun. Nonlinear Sci. Numer. Simul.* **2016**, *39*, 252–270. [CrossRef]
26. Hartigan, J.A. *Clustering Algorithms*; John Wiley & Sons: New York, NY, USA, 1975.
27. Shepard, R.N. The analysis of proximities: Multidimensional scaling with an unknown distance function. *Psychometrika* **1962**, *27*, 125–140. [CrossRef]
28. Kruskal, J. Multidimensional scaling by optimizing goodness of fit to a nonmetric hypothesis. *Psychometrika* **1964**, *29*, 1–27. [CrossRef]
29. Székely, G.J.; Rizzo, M.L. Hierarchical Clustering via Joint Between-Within Distances: Extending Ward's Minimum Variance Method. *J. Classif.* **2005**, *22*, 151–183. [CrossRef]
30. Machado, J.A.T. Visualizing Non-Linear Control System Performance by Means of Multidimensional Scaling. *ASME Comput. Nonlinear Dyn.* **2013**, *8*, 041017. [CrossRef]
31. Deza, M.M.; Deza, E. *Encyclopedia of Distances*; Springer: Berlin/Heidelberg, Germany, 2009.
32. Machado, J.A.T.; Lopes, A.M. Multidimensional Scaling Locus of Memristor and Fractional Elements. *J. Adv. Res.* **2020**, *25*, 147–157. [CrossRef] [PubMed]
33. Machado, J.A.T.; Lopes, A.M. newblock Multidimensional Scaling and Visualization of Patterns in Prime Numbers. *Commun. Nonlinear Sci. Numer. Simul.* **2020**, *83*, 105128. [CrossRef]
34. Marvasti, F. A new method to compensate for the sample-and-hold distortion. *IEEE Trans. Acoust. Speech Signal Process.* **1985**, *33*, 738–741. [CrossRef]
35. Wang, Z. The fastest sample-and-hold circuit. *Microelectron. J.* **1990**, *21*, 49–52. [CrossRef]

36. Van De Plassche, R. Sample-and-hold amplifiers. In *Integrated Analog-To-Digital and Digital-To-Analog Converters*; Springer: Boston, MA, USA, 1994; p. 264. [CrossRef]
37. Taylor, H.R. Sample and hold circuits. In *Data Acquisition for Sensor Systems*; Springer: Boston, MA, USA, 1997. [CrossRef]
38. Abolhasani, A.; Tohidi, M.; Hadidi, K.; Khoei, A. A new high-speed, high-resolution open-loop CMOS sample and hold. *Analog. Integr. Circuits Signal Process.* **2014**, *78*, 409–419. [CrossRef]
39. Psychalinos, C.; Elwakil, A.; Maundy, B.; Allagui, A. Analysis and realization of a switched fractional-order-capacitor integrator. *Int. J. Circuit Theory Appl.* **2016**, *44*, 2035–2040. [CrossRef]
40. Pelgrom, M. *Analog-to-Digital Conversion*; Springer: Cham, Switzerland, 2017.
41. Yuz, J.I., Goodwin, G.C.; Garnier, H. Generalised hold functions for fast sampling rates. In Proceedings of the 43rd IEEE Conference on Decision and Control, Atlantis, Bahamas, 14–17 December 2004; Volume 2, pp. 908–1913. [CrossRef]
42. Trivedi, R. Low Power and High Speed Sample-and-Hold Circuit. In Proceedings of the 2006 49th IEEE International Midwest Symposium on Circuits and Systems, San Juan, PR, USA, 6–9 August 2006; pp. 453–456. [CrossRef]
43. De la Sen, M. About Optimal Fractional Hold Circuits for Inter-sample Output Reconstruction in Sampled-data Systems. *Sensors* **2007**, *7*, 3146–3155. [CrossRef] [PubMed]
44. Nishi, M.; Ishitobi, M. Sampled-Data Models for Nonlinear Systems with a Fractional-Order Hold. In Proceedings of the 18th IEEE International Conference on Control Applications Part of 2009 IEEE Multi-Conference on Systems and Control, Saint Petersburg, Russia, 8–10 July 2009; pp. 153–158. [CrossRef]
45. Nishi, M.; Ishitobi, M. Sampled-data models for affine nonlinear systems using a fractional-order hold and their zero dynamics. *Artif. Life Robot.* **2010**, *15*, 500–503. [CrossRef]
46. Zeng, C.; Xiang, S.; He, Y.; Ding, Q. Nonlinear Sampled-Data Systems with a Generalized Hold Polynomial-Function for Fast Sampling Rates. *J. Syst. Sci. Complex.* **2019**, *32*, 1572–1596. [CrossRef]

Publisher's Note: MDPI stays neutral with regard to jurisdictional claims in published maps and institutional affiliations.

applied sciences

MDPI

Article

Electro-Thermal and Aging Lithium-Ion Cell Modelling with Application to Optimal Battery Charging

Sara Mohajer [1,2], Jocelyn Sabatier [2,*], Patrick Lanusse [3] and Olivier Cois [1]

1 Robert Bosch GmbH, Wernerstraße 51, 70469 Stuttgart, Germany; Sara.Mohajer@de.bosch.com (S.M.); Olivier.Cois@de.bosch.com (O.C.)
2 IMS Lab., Bordeaux University, UMR 5218 CNRS, 351 Cours de la Libération, 33405 Talence, France
3 IMS Lab., Bordeaux INP, UMR 5218 CNRS, 351 Cours de la Libération, 33405 Talence, France; patrick.lanusse@ims-bordeaux.fr
* Correspondence: Jocelyn.sabatier@u-bordeaux.fr

Received: 1 May 2020; Accepted: 7 June 2020; Published: 11 June 2020

Abstract: This paper deals with optimal charging versus aging minimization for lithium-ion batteries. The optimal charging strategy proposed involves charging controllers whose design relies on a battery model. The model, especially designed for automotive battery management systems applications, is recalled in this paper. It provides the voltage response of a cell to an input current. It also models side reactions that produce degradation mechanisms and thus decrease battery performance. Side reaction modelling involves taking into account the temperature cell variations, which are thus also modelled. The association of the three above-mentioned sub-models leads to an electro-thermal battery aging model used to design an optimal charging strategy that simultaneously takes into account the minimization of charging time and maximization of battery lifetime. Thus, to achieve a charging controller that manages battery health, an appropriate charging trajectory was computed by solving an optimization problem minimizing aging. Then, a charge control loop was designed. The nonlinear behavior of the battery was taken into account through the linearization of the electro-thermal aging model in different operating conditions. To take into account the resulting linear model family, the CRONE design methodology was used. The principles of this methodology are recapped and the design of the charging control loop is explained. The efficiency of the resulting charge controller is illustrated by several simulations.

Keywords: lithium-ion batteries; fast charging; battery ageing

1. Introduction

A long lifetime is a very important requirement for the design of electric and hybrid-electric vehicles. To meet this requirement, battery lifetime must be optimized. Discharge phases are not considered here, as they mainly depend on user demands. However, during charging it is simultaneously possible to optimize the charging time and the battery degradation. It is crucial to monitor battery degradation during charging, as hard charging profiles are usually used to quickly arrive at the desired State of Charge (SOC). With such charging profiles, battery internal temperatures, and thus the rate of unwanted side reactions, increase. Lithium ions are consumed by side reactions and can thus no longer participate in the intercalation process. This process causes the aging of the battery and hence poor performance.

A fast charging controller can be implemented to deal with this problem. Using information on the battery state, through direct measurement and using a model for some internal variables that cannot be measured, the charging controller can be used to adjust operating limitation to prevent ageing.

The battery model solves the compromise between accuracy (predominant electrochemical phenomena must be taken into account) and parsimony (efforts for model parameterization and simulation must be limited).

Such a compromise can be reached using the single particle fractional model that was introduced in [1], or in [2,3] for simplified versions. These models involve fractional transfer functions that are particularly efficient at modelling the diffusion phenomena that take place in batteries [2] and also in several other electrochemical devices. Several papers have demonstrated the efficiency of these models [3–5], particularly when they are used for charge estimation. Extensions of these models to take into account cell aging and cell thermal behaviour can be found in [6]. Compared to other models proposed in the literature, the model in [6] is a physics-based model that takes into account degradation mechanisms in terms of temperature, SOC and C-rate (discharge or charge current normalised by the cell capacity), thus capturing experimental measures well.

Several control strategies have been proposed for fast charging control taking battery state into account. The strategies can be classified into two categories: open-loop and closed-loop strategies.

Model based open-loop optimal control has been considered in several papers. In [7], a Taguchi type algorithm was used to compute an optimal charging profile that minimises a cost function involving energy efficiency, charging time and temperature variation. In this approach, the model used is an Equivalent Circuit Model (ECM), which does not take cell degradations into account. In [8], an optimal open loop current profile, analytically defined, was computed using an optimization problem that minimizes a multi-objective function involving energy loss, temperature rise and charging time to charge. Such an approach is also based on an ECM. In [9], a more complex battery ECM that takes into account aging and thermal behavior was implemented. It was used to solve an open-loop multi-objective optimization problem based on a weighted sum of charging time and aging.

The open-loop approaches previously discussed suffer from a high sensitivity to parametric uncertainties coming from the approximations made to derive a battery model. To overcome these modelling errors, a closed loop approach is required.

Some closed-loop optimal charging strategies have been proposed in the literature. In [10], charging time was minimised using a feedback controller. Excessive anode surface concentration and the increase of temperature were assumed to be the reason for aging, and were thus considered as optimisation constraints. A physics-based electro-thermal model was used for the battery state estimation, but suffered from a poor state model state observability (mainly for the estimation of the electrodes' solid-phase concentrations).

Extensive work has been performed to address optimal charging for aging minimization with Model Predictive Control (MPC) for various application fields. An aging-aware predictive control strategy of PV-battery was presented in [11]. In this work, a convex MPC problem was solved to seek the optimal balance between the building utility cost and the battery life-cycle cost. In [12], a MPC scheme based on a linearized version of a Pseudo-two Dimensional (P2D) model was proposed in order to track a reference value of the SOC, while taking into account the aging dynamics of the system, as well as temperature and voltage constraints. Ref. [13] presents an MPC application to minimize the charging time of a lithium-ion battery subject to electrochemical and thermal constraints. The satisfaction of these constraints ensures that the battery degradation is minimized, or at least mitigated. A new methodology for battery charging control enabling an optimal tradeoff between the charging time and battery state-of-health (SOH) was also proposed in [14]. Using a model reduction approach, a physics-based low-order battery model was first proposed and used to formulate a model-based charging strategy. The optimal fast charging problem was formulated in the framework of tracking MPC.

In [15], a state-feedback nonlinear MPC approach was introduced, with the objective of minimising charging time. Constraints on the temperature and on the Solid Electrolyte Interphase (SEI) overpotential (assumed to be the causes of aging) were also taken into account. An MPC strategy was also used in [16] for charging control. A genetic algorithm was combined with MPC for charging time minimisation and to keep the battery temperature above a fixed limit to reduce aging. In [17], a nonlinear MPC was used for the energy management of a power-split hybrid electric vehicle to improve battery aging while maintaining the fuel economy at a reasonable level.

In all these closed loop charging strategies, in spite of their valuable contribution, the MPC approach does not ensure robustness with respect to rapid variations in state variables (for instance, anode potential). As a battery is a highly nonlinear system, any linear approximation depends strongly on the signal levels (states and inputs). With MPC, the model parameters can be time-varying. However, the control design is based on the nominal plant and needs information on the system disturbances and on the exact dynamic behaviour in each control sequence. This leads to a lack of efficiency in relation to uncertainties and disturbances.

Fast charging closed loop strategies also suffer from another defect. No study considers the battery aging produced by the magnitude of the charging current applied. This magnitude is, however, known to greatly aggravate battery aging.

Referring to this state of the art in the field of lithium-ion fast charging strategies and using a physics-based model of the cell, this paper addresses the following main challenges:

- the design of a charging strategy that permits fast charging and a minimisation of aging resulting from battery temperature increase and large current magnitude,
- the robustness of the strategy with respect to large dynamic behaviour variations.

The paper is organized as follows. Section 2 describes the electro-thermal aging model that will be used in the sequel for the design of an optimal fast charging strategy. The optimal charging problem for a Li-ion battery of PHEVs (Plug-in Hybrid Electric Vehicles) is formulated in Section 3. An example of an optimal current profile and the corresponding battery model response are described in Section 4. The optimal charging problem formulated in Section 3 is solved in Section 5 with two imbricated control loops that involve robust controllers. The design of these controllers is based on a family of models that capture the nonlinear behaviour of the cell around several operating points. This family of models is computed in Section 4.2.

2. Cell Modelling

A detailed description of the lithium-ion cell model used in this work is given in [6]. It comprises three parts:

- an electrochemical part,
- a thermal part,
- an aging part.

The validation of this model and of all its parts is proposed in [6]. This paper only gives the main equations.

2.1. Description of the Electrochemical Part of the Model

The equations given in [18,19] were used to define the electrochemical part of the model. These equations result in Newman's modelling approach [20], which leads to a pseudo 2D model and assumes that each electrode of a cell is replaced by a single spherical particle in which the Li+ ions are inserted. In the resulting particle, and through several simplifications that are given in [3], the lithium ions' concentration gradient is described by the following diffusion equation:

$$\frac{\partial c_s}{\partial t} = \frac{D_s}{r^2}\frac{\partial}{\partial r}\left(r^2\frac{\partial c_s}{\partial r}\right)\left\{\begin{array}{ll} \left.\frac{\partial c_s}{\partial t}\right|_{r=0} & = 0 \\ D_s\left.\frac{\partial c_s}{\partial r}\right|_{r=R_s} & = -\frac{j^{Li}_{mean}}{a_s F} \end{array}\right. \tag{1}$$

In Relation (1), J_{mean}^{Li} is the average of the electrode current density, r is the sphere radius and D_s is the diffusion coefficient. Resolution of the differential Equation (1) leads to the following transfer function linking the lithium concentration at the surface of the particle denoted $C_s(r = R_s)$ to J_{mean}^{Li}:

$$G(s) = \frac{C_s(r = R_s)}{J_{mean}^{Li}} = -\frac{R_S}{FD_s a_s\left(\sqrt{\frac{s}{D_s}}\coth\left(\sqrt{\frac{s}{D_s}}R_S\right) - 1\right)} \tag{2}$$

The transfer function (2) cannot be easily simulated due to the coth function. As demonstrated in [3], it can be approximated by the transfer

$$H(s) = \frac{K_1}{s}\left(1 + \frac{s}{\omega_{cs}}\right)^{0.5} \tag{3}$$

where the parameters K_1 and ω_{cs} are functions of the coefficients of the diffusion Equation (1) (see [3]).

For a consistent initialization of the model, and also to make the variable SOC appear, the transfer function $H(s)$ can be split into two transfer functions, such that

$$H(s) = H_{avg}(s) + H_{part}(s) \tag{4}$$

with

$$H_{avg}(s) = \frac{C_{s,avg}}{j_{mean}^{Li}} = \frac{K_1}{s} \tag{5}$$

and

$$H_{part}(s) = \frac{\Delta C_s}{j_{mean}^{Li}} = H(s) - H_{avg}(s) = \frac{K_1}{s}\left(\left(1 + \frac{s}{\omega_{cs}}\right)^{0.5} - 1\right) \tag{6}$$

In Relation (4), $H_{avg}(s)$ represents the average concentration (from the centre to the surface of the particle) and $H_{part}(s)$ is the partial gradient of ion concentration on the particle surface.

For the implementation of the battery model in a vehicle control unit, a discrete form of the transfer functions $H_{avg}(s)$ and $H_{part}(s)$ can be computed, as shown in [6]. Also, a polynomial can be used to link the lithium concentration at the surface of the particles, denoted SOC_{surf}, to the cell Open Circuit Potential (OCP), denoted U_n and U_p respectively for the anode and the cathode. If the charge transfer resistance is denoted R_{ct} (R_{ct} can be deduced from the Butler–Volmer equation linearization), the resulting electrochemical sub-model is represented by Figure 1.

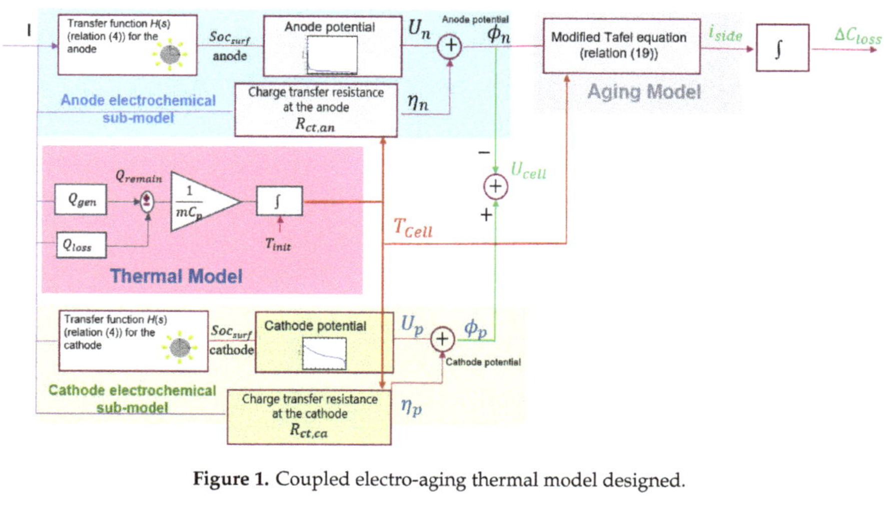

Figure 1. Coupled electro-aging thermal model designed.

2.2. Description of the Thermal Part of the Model

The kinetics of charge transfer during the intercalation process depends in particular on cell temperature. Cell temperature also has an impact on the kinetics of the side reactions. It is thus essential to associate a thermal model to the electrochemical sub-model. In this work, the efficient and simple thermal model presented in [21] was used to estimate the cell temperature. This thermal model results from a heat balance within the cell, and is described by

$$mC_p \frac{dT(t)}{dt} = Q_{gen}(t) - Q_{loss}(t) \tag{7}$$

The thermal parameters associated to Relation (7) are given in Table 1. The generated heat is defined by

$$Q_{gen}(t) = U_{pol}(t)I(t) + R_f I(t)^2 \tag{8}$$

In Relation (8), $U_{pol}(t)$ is the irreversible heat generation due to electrode polarization. $R_f I(t)^2$ models heat losses over the internal resistor of R_f.

Table 1. Parameters of the thermal model.

Symbol	Parameter	Unit
m	Mass of the cell	kg
C_p	Specific heat capacity	$J{\cdot}kg^{-1}{\cdot}K^{-1}$
U_{pol}	Polarization voltage	V
R_f	High frequency resistance	Ω
I	Input current	A
α	Heat transfer coefficient	$W{\cdot}m^{-2}{\cdot}K$
A	Cell surface area	m^2
T_{amb}	Ambient temperature	K

The convective heat exchanged with the environment is:

$$Q_{loss}(t) = \alpha A(T(t) - T_{amb}) \tag{9}$$

2.3. Description of the Aging Part of the Model

Depending on the intended application, fast charging or battery cycling until End of Life (EOL), a short-term or a long-term model is required. In this work, the developed model is devoted to designing fast charging algorithms, and aging is considered to be mainly caused by the growth of a Solid Electrolyte Interphase (SEI) layer on the anode. This degradation mostly happens during short to middle-term battery usage during fast charging cycles. The aging model used is based on the equations proposed in [22] and, following the observations from post-mortem analysis of the investigated cells, aging is only considered on the anode electrode. For a constant temperature T, the current produced by the side reaction leading to the undesired reduction of Li ions is defined by the Tafel equation

$$j_{sr} = -a_s^n j_{0,sei} exp\left(-\frac{nF\alpha_n}{RT}\eta_{sei}\right) \tag{10}$$

If φ_{en}, R_{sei} and U_{sei}^{ref} denote respectively the anode potential, the inner resistance of the SEI layer and the reference potential of SEI formation (assumed to be 0.4 V), the SEI overpotential is given by the relation

$$\eta_{sei} = \varphi_{en} - R_{sei}I - U_{sei}^{ref} \tag{11}$$

in which the anode potential φ_{en} is defined by

$$\varphi_{en} = \eta_n + U_n + R_{sei} I \tag{12}$$

According to [23], η_n is deduced from the Butler–Volmer equation which defines the kinetics of Li ion intercalation, and can be approximated by

$$\eta_n = \frac{2RT}{F} \log\left(\left(\frac{j_I}{2a_s^n j_0^n}\right) + \sqrt{\left(\frac{j_I}{2a_s^n j_0^n}\right)^2 + 1}\right) \tag{13}$$

for an easier implementation. The electrochemical parameters associated with Relations (10) to (13) are defined in Table 2. Some parameters are taken from [22]. The others were estimated using measurements on the cell under investigation.

The increase in temperature resulting from fast charging operation has a large impact on the side reaction rate. Thus, Relation (10) must be modified to take into account the Arrhenius dependency of the side reaction current on the temperature variation. The modified Equation (10) is thus given by

$$j_{sr} = -a_s^n j_{0,sei} exp\left(\frac{E_a}{R}\left(\frac{1}{T_{ref}} - \frac{1}{T}\right)\right) exp\left(-\frac{nF\alpha_n}{RT}\eta_{sei}\right) \tag{14}$$

The SEI thickness (δ_{sei}) and resistance (R_{sei}) can be computed by the following Relations [22]:

$$\frac{\partial \delta_{sei}}{\partial t} = \frac{M_{sei}}{2F\rho} j_{sr} \tag{15}$$

$$R_p(t) = \frac{\delta_{sei}}{\kappa_{sei}} \tag{16}$$

$$R_{sei}(t) = R_{sei,init} + R_p(t) \tag{17}$$

Table 2. Aging sub-model parameters.

Symbol	Parameter	Unit
a_s^n	Specific surface area	m^{-1}
$j_{0,sei}$	Exchange current density	$A \cdot m^{-2}$
α_n	Symmetry factor	-
n	Number of transferred electrons	-
E_a	Activation Energy	$kJ \cdot mol^{-1}$
δ_{sei}	SEI layer thickness	m
M_{sei}	Molar mass of SEI layer	$kg \cdot mol^{-1}$
ρ	Density of SEI layer	$kg \cdot m^{-3}$
κ_{sei}	SEI layer conductivity	$S \cdot m^{-1}$
$R_{sei,init}$	Initial resistance of SEI layer	$\Omega \cdot m^2$
R_p	Resistance of side reaction product	$\Omega \cdot m^2$
I_{ch}	Charging current	A

Experimental observations [24] have shown that, after a primary formation and after some cycles, the SEI layer thickness growth at the surface of the electrode decreases. To model such a behavior, an exponential function was associated to Relation (18), which becomes

$$j_{sr} = -a_s^n j_{0,sei} exp(-\lambda \delta_{sei}) exp\left(\frac{E_a}{R}\left(\frac{1}{T_{ref}} - \frac{1}{T}\right)\right) exp\left(-\frac{nF\alpha_n}{RT}\eta_{sei}\right) \tag{18}$$

where λ is the SEI decay rate constant (m^{-1}). To obtain this coefficient, it is assumed that, after some cycles, the SEI thickness stabilizes.

To guarantee an efficient fast charging strategy, the aging sub-model must also reproduce with good accuracy the higher capacity loss that has been observed at higher charging C-rates. This requirement is not permitted with Relation (18). Thus, to scale the inverse Butler-Volmer overpotential η_n, a coefficient K_η is introduced in Equation (18), which thus becomes

$$j_{sr} = -a_s^n j_{0,sei} exp(-\lambda \delta_{sei}) exp\left(\frac{E_a}{R}\left(\frac{1}{T_{ref}} - \frac{1}{T}\right)\right) exp\left(\frac{-nF\alpha_n}{RT}\left(K_\eta \eta_n + U_n - U_{sei}^{ref}\right)\right) \quad (19)$$

with $C_T = -a_s^n j_{0,sei} exp\left(\frac{nF\alpha_n}{RT} U_{sei}^{ref}\right) exp\left(\frac{E_a}{RT_{ref}}\right)$.

In Relation (19), it can be noticed that the time-dependent (calendar) aging effects and current-dependent (cycling) aging effects are decoupled. This offers the possibility of better interpreting the influence of each mechanism on cell aging, and of emphasizing the influence of the C-rate on the decrease in capacity during cycling and in particular rapid charges.

To the best knowledge of the authors, the physics-based battery model proposed in this study, which takes into account degradation mechanisms as a function of SOC, temperature and C rate, is unprecedented in the literature. These characteristics also allow a very good correspondence between the response of the model and the measurements.

2.4. Whole Cell Model

The electro-thermal aging model described in the previous paragraphs can be represented by the block diagram of Figure 1. In this figure, the sub-models denoted 'Anode' and 'Cathode' are deduced from the description performed in Section 2.1. The thermal model of Section 2.2 provides the cell temperature information from the current applied to the aging model described in Section 2.3. This model has been validated by several tests, as described in [6].

Regarding the tuning of the parameters associated with this model, it can be partially accomplished using information provided by the battery manufacturer. Tests are required to tune the remaining parameters. The tuning methodology is described in [5], and needs to be applied for each new type of battery. As no feedback exists between the aging model and the electrochemical sub-models, the model can fit the behaviour of the battery only within a given aging range (this is enough to get the battery state estimation during a charge). The Model of Figure 1 must be adjusted (on board) periodically to take into account the impact of aging on some parameters of these sub-models (OCP functions, charge transfer resistance, ...).

Further studies can be conducted (see [3]) to include the electrolyte dynamics in the anode and cathode electrochemical sub-models. Considering the ion diffusion phenomenon inside the electrolyte would increase the model accuracy in estimation of the terminal voltage, but would also increase the model's complexity.

3. Trajectory Planning

The fast charging strategy developed in this paper, which can be integrated in a battery management system (BMS), is described globally by Figure 2. In this diagram, the model presented in Section 2 is used in the battery observer to simulate the battery behavior under various operating conditions. Using measures of the cell current I_{Charge}, voltage U_{cell} and temperature T_{cell}, this observer estimates (among other information) the cell capacity fade ΔC_{fade}. This information is used by a fast charging controller that adapts the charge current magnitude to follow a charge profile I_{ref} while reducing the cell capacity fade during recharge.

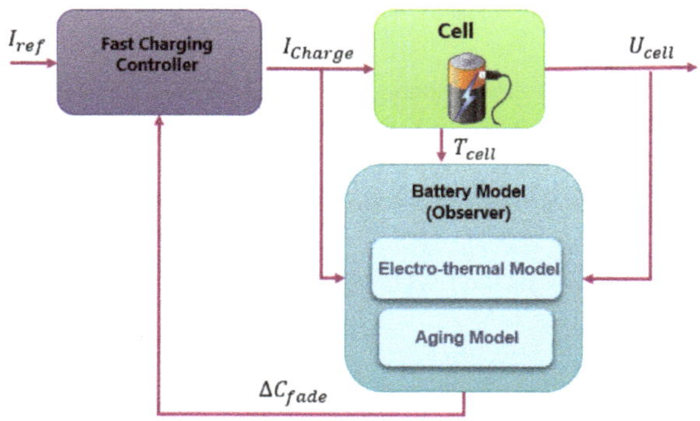

Figure 2. Description of the implementation of the fast charging functions in a battery management system (BMS).

The function of the charging controller is to determine the optimal charging current profile imposed on the battery. The optimal current is that which minimizes the recharging time (therefore the highest possible) while limiting a variable upper limit on the current. This makes it possible to meet a rapid recharging target without increasing battery aging too much. The strategy adopted in this work to determine the optimal charging profile thus consists in computing offline the side reaction current before battery charging. The side reaction current changes over time and is a function of the SOC and of the charging current. If the charging process is split into a finite number of time slots, a lookup table can provide the limit on the side reaction current on each interval as a function of SOC. The side reaction current limit provided on each slot defines the side reaction current trajectory. The controller thus computes the corresponding charging current and defines in real time the optimal profile of the charging current to meet the calculated side reaction current trajectory.

In practice, to acheive the optimal charging trajectory, the capacity loss during a fixed charging time t_f of twenty minutes is minimised under the following constraints:

- bounds on the charging current (Ich), in order to avoid exceeding the maximum charge current limits (3.5 C) for safety reasons,
- an increase in the SOC from 5% to 80% (but other ranges of SOC variation can be defined).

Denoted in the sequel is a Multi-stage Constant Current (MCC) of duration t_f, a charge current profile made of a series of N pulses (or slots) of the same duration (t_f/N), in which each pulse can have a different magnitude (thus N degree of freedom in the current profile that can be optimised). An example of an MCC profile is represented in Figure 3. It is used for its flexibility in the definition of a profile in order to maximize energy efficiency and to improve battery life during fast charging [25]. The optimal control problem can thus be analytically defined by

$$\min_{I_{ch}(t)} \int_0^{t_f} \Delta C_{loss}(t)dt \tag{20}$$

subject to

$$\dot{x}(t) = f(x(t), I_{ch}(t)), x(0) = x_0 \tag{21}$$

$$\frac{1}{3600}\int_0^{t_f} I_{ch}(t)dt \geq Q_{trans} \tag{22}$$

$$0 \leq I_{ch}(t) \leq 3.5C \tag{23}$$

in which Q_{trans} denotes the Li-ion charges stored in (Ah), ΔC_{loss} denotes the capacity loss (Ah), and t_f (s) is the charging time. Sequential Quadratic Programming (SQP) is used to solve the optimal control problem defined by Relations (20) to (23) for various operating conditions.

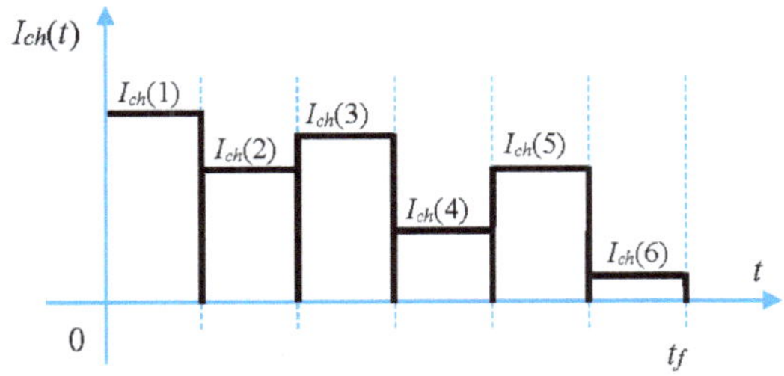

Figure 3. An example of a Multi-stage Constant Current (MCC) profile with 6 slots ($N = 6$). Each value $I_{ch}(k)$ has to be optimised to find the current optimal profile.

Variations of SOC, anode potential, side reaction and capacity loss variation at 35 °C obtained after optimisation with only 4 slots are shown in Figure 4. These curves are compared to those obtained with a conventional Constant Current Constant Voltage (CCCV) profile. This kind of charge profile has been chosen for comparison purposes because it is very commonly used for recharging batteries (lithium ion or not). This figure shows a decrease in capacity loss close to 15% with the optimal current profile in comparison with the CCCV profile. The optimisation was performed for various aging levels, various ambient temperatures (10, 17, 27, 30, 47, and 77 °C) and for various states of health from End of Life (EOL), Middle of Life (MOL) and Beginning of Life (BOL).

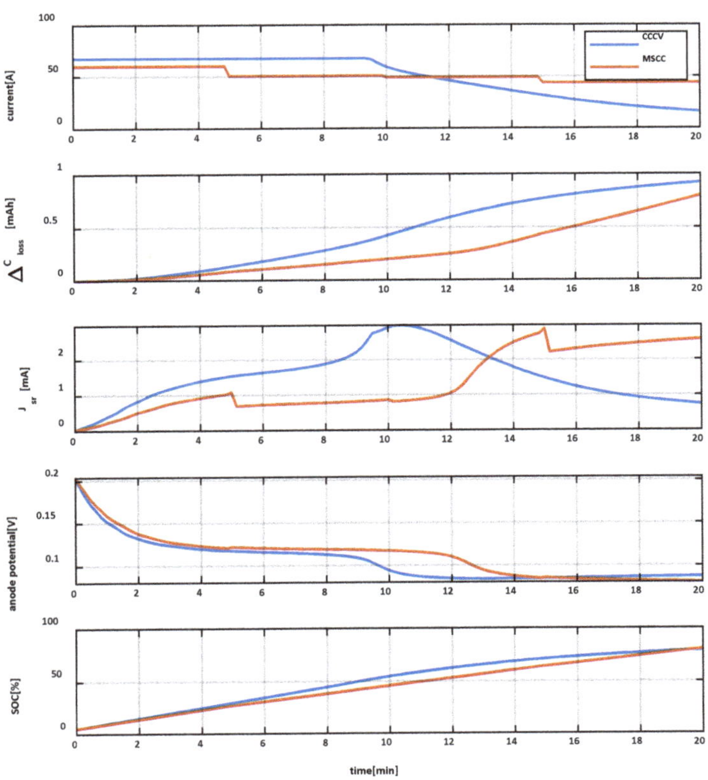

Figure 4. Trajectories computed for a fixed charging time of 20 min, at T = 35 °C, for a cell at the Beginning of Life (BOL) in a State of Charge (SOC) range of 5–80%.

These results were obtained with a cell whose dynamical behaviour is assumed to be perfectly known. This is an ideal case that does not exist in practice due to modelling uncertainties, and applying the optimal charging current profiles obtained does not necessarily minimize aging while reaching the target SOC. To take into account the modelling uncertainties, a closed loop strategy must be implemented. As shown in the next section, the side reaction current J_{sr} that creates aging was defined as the control signal to control the battery charging due to its high sensitivity to the charging current, as illustrated by Figure 4.

4. Cell Model Linearization

The interest of closed-loop control for fast charging has been highlighted in Sections 1 and 3. The design of a linear controller to implement such a loop is thus the objective of this section. The block diagram of the cell model in Figure 1 suggests that the cell is highly nonlinear. This is confirmed by the simulations of this model, which show a high sensitivity to temperature, lifetime stage and charging current level. Such a nonlinear behaviour is taken into account in the approach presented here with a linearized model set that is computed for a set of operating points (temperature, charging current, stage of life) and various optimal charging profiles, and thus side reaction current (J_{sr}) profiles.

4.1. State-Space Model of the Cell Model

The following nonlinear state-space model:

$$\begin{cases} \dot{x}(t) = f(x(t),\ u(t)) \\ y(t) = g(x(t),\ u(t)) \end{cases} \tag{24}$$

represents the battery model in which the input $u(t)$ is the charging current $I_{ch}(t)$. The state vector $x(t)$ and the output vector $y(t)$ are respectively given by

$$y = [J_{sr}, U_{bat}, \phi_{en}, OCV,\ T, SOC]^T \tag{25}$$

$$x = \left[\Delta C_{loss}, \delta_{SEI}, C_{avg,n}, C_1^{part,n} \ldots C_5^{part,n}, C_{avg,p}, C_1^{part,p} \ldots C_5^{part,p}\right]^T \tag{26}$$

where the variables that appear in Relations (25) and (26) are defined in Table 3.

Table 3. Variables in the state and output vectors of the nonlinear cell model * I = p (cathode), n (anode).

Symbol	Parameter	Unit
ΔC_{loss}	Capacity loss	mAh
δ_{SEI}	Thickness of SEI layer	m
$C_{avg,i}$ *	Electrode average concentration	Ah
$C_1^{part,i}$ *	Electrode partial concentration	Ah
U_{bat}	Battery terminal voltage	V
ϕ_{en}	Anode potential	V
OCV	Open Circuit Voltage	V

The next section describes the approach used to derive a set of linear models (one model for a given operating point) for the approximation of the nonlinear cell model.

4.2. Operating Points Definition

The optimal charging profiles computed in Section 3 were used to define the operating points required for the linearization of the nonlinear model (24). The optimization was performed for various aging levels and ambient temperatures, with SOC within 5–80%. For all optimal current profiles, relaxation intervals were added to reach equilibrium conditions during the constant current phases of the MCC profile. 432 operating points were finally considered, made up of 18 ambient (and internal) temperatures, three aging levels and eight operating conditions for each charging profile.

4.3. Uncertain Linear Models Resulting from the Nonlinear Battery Model

The nonlinear model (24) was linearized around operating points using a numerical method (Matlab linmod command). The Bode diagrams of the resulting set of linear models of the transfer function $J_{sr}(s)/I_{ch}(s)$ are shown in Figure 5. Among this set, one was chosen as a nominal model, with the other linear model thus defining an uncertain linear model. Figure 5 highlights a strong dependence of the frequency response of the transfer function $J_{sr}(s)/I_{ch}(s)$ to the operating point. The resulting gain and phase uncertainty are taken into account in the next section for the design of a fast charging robust controller.

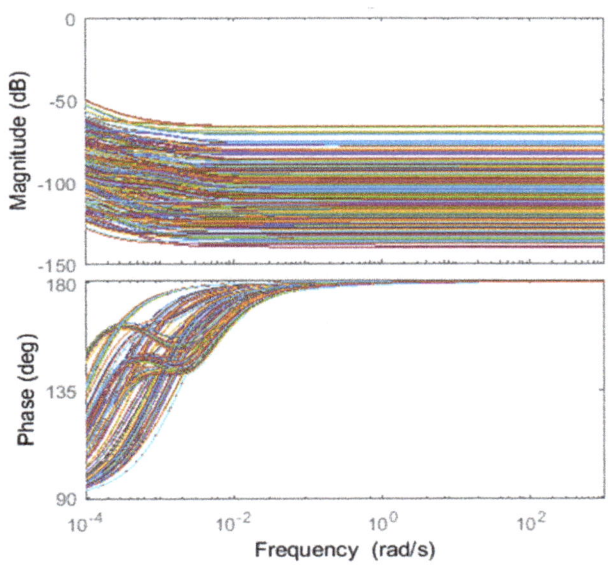

Figure 5. Tansfer function $J_{sr}(s)/I_{ch}(s)$ frequency responses for the considered operating points.

5. Design of the Fast Charging Robust Controller

5.1. Closed-Loop Control

The objective was to design a two-degrees of freedom fast charging controller, as illustrated by Figure 5. In such a control scheme, the feedforward must ensure the initialization of the plant input, and the robust feedback must permit the desired accurate output tracking. The plant model output $J_{sr,traj}$ and input $I_{Ch,traj}$ reference signals (optimal trajectories) are known and provided by the optimal MCC trajectory planning.

As shown in Figure 6, a low pass filter was implemented for the reference current profile smoothing. With the control scheme of Figure 6, if the battery behaviour fits the one used for the charging profile optimisation, the charging current I_{Charge} that makes it possible to track the optimal $J_{sr,traj}$ is equal to the feedforward current $I_{Ch,FF}$. The feedback controller thus has no effect. Conversely, if the battery behaviour differs from the behaviour of the model used for the optimisation of the charging profiles, the feedback loop adapts the charging current. It imposes the computed optimal side reaction current on the cell. As only cell temperature T, charging current I_{Charge} and voltage U_{cell}, are measureable signals, an aging observer using the cell model was designed to estimate the side reaction current.

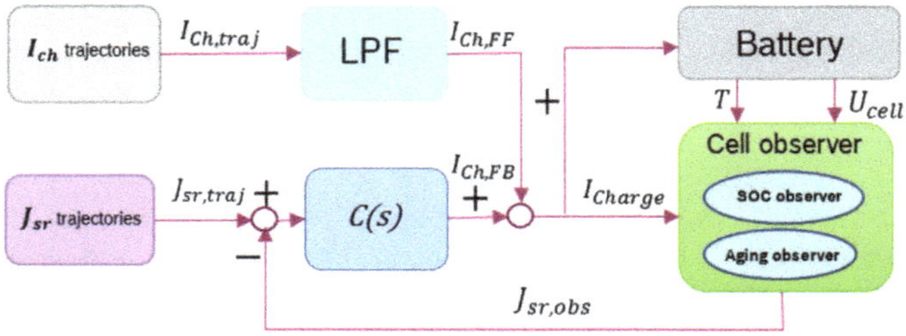

Figure 6. Fast charging control closed-loop.

5.2. CRONE Control Methodology for Robust Controller Design

The controller $C(s)$ in Figure 6 was designed using the CRONE control methodology. It uses fractional differentiation to define high-level design parameters [26] that are tuned in the frequency domain. The third generation CRONE control was used in this work. It is based on the optimization of the transfer function $\beta_{nom}(s)$ parameters ($\beta_{nom}(s)$ being computed for the nominal plant denoted G_{nom}) defined by

$$\beta_{nom}(s) = \beta_l(s)\beta_m(s)\beta_h(s) \tag{27}$$

in which $\beta_l(s)$ is an integer proportional integrator of order n_l defined by

$$\beta_l(s) = C_l\left(\frac{\omega_{N^-}}{s} + 1\right)^{n_l} \tag{28}$$

$\beta_m(s)$ groups several band-limited generalized templates

$$\beta_m(s) = \prod_{k=-N^-}^{N^+} \beta_{mk}(s) \tag{29}$$

with

$$\beta_{mk}(s) = C_k\left(\alpha_k\frac{1+\frac{s}{\omega_{k+1}}}{1+\frac{s}{\omega_k}}\right)^{a_k}\left(\Re_i\left\{\left(\alpha_k\frac{1+\frac{s}{\omega_{k+1}}}{1+\frac{s}{\omega_k}}\right)^{ib_k}\right\}\right)^{-q_k sign(b_k)} \tag{30}$$

where $\alpha_k = \left(\frac{\omega_{k+1}}{\omega_k}\right)^{0.5}$ for $k \neq 0$, and $\alpha_0 = \left(\frac{1+\left(\frac{\omega_r}{\omega_0}\right)^2}{1+\left(\frac{\omega_r}{\omega_1}\right)^2}\right)^{0.5}$ and $\beta_h(s)$ is a low-pass filter of order n_h:

$$\beta_h(s) = C_h\left(\frac{s}{\omega_{N^+}} + 1\right)^{-n_h} \tag{31}$$

Order n_l must be adjusted to get the desired controller accuracy. The constants C_l, C_k and C_h are tuned to ensure a given resonance frequency ω_r of the closed loop transfer function $T_{nom}(s) = \beta_{nom}(s)/(1 + \beta_{nom}(s))$. Order n_h was chosen to ensure a bi-proper or strictly proper controller. The open-loop parameters are tuned M_T, where M_T denotes the resonance peak of the complementary sensitivity function $T(s)$. Such an objective is attained by minimizing the objective function

$$J = \sup_G M_T - M_{Tnom} \tag{32}$$

in which M_{Tnom} denotes a whised value for the closed-loop resonance peak computed with the nominal plant denoted G_{nom}. The minimisation of Relation (32) is achieved under the constraints on the four closed loop sensitivity functions that follows:

$$\inf_G|T(j\omega)| \geq T_l(\omega),\ \sup_G|T(j\omega)| \leq T_u(\omega) \sup_G|S(j\omega)| \leq S_u(\omega),\ \sup_G|CS(j\omega)| \leq CS_u(\omega) \sup_G|GS(j\omega)| \leq GS_u(\omega) \quad (33)$$

with

$$\begin{cases} T(s) = \frac{G(s)C(s)}{1+G(s)C(s)}\ S(s) = \frac{1}{1+G(s)C(s)} \\ CS(s) = \frac{C(s)}{1+G(s)C(s)}\ SG(s) = \frac{G(s)}{1+G(s)C(s)} \end{cases} \quad (34)$$

Using the Nichols plot, by minimizing criterion J in Relation (32), the optimal parameters move the frequency uncertainty domains so that they cover areas of low stability margin as little as possible around the (−180, 0 dB) critical point on the Nichols chart (as the open-loop frequency response uncertainty domains are those of the plant). Only a nonlinear optimization method can be used to compute the optimal parameters as uncertainties are not norm-bounded and taken into account by the least conservative method. If $N^+ = N^- = 0$ in Relation (29), only the independent four parameters ω_0, ω_1, ω_r and $Y_r = |\beta(j\omega_r)|_{dB}$ need to be optimized in Relation (27). In the Nichols chart, the tangency of the frequency response $\beta_{nom}(j\omega)$ to the desired M_{Tnom} circle can be ensured by computing the other parameters of Relation (27), such as a_0 and b_0. The optimal open loop transfer function being found, the controller $C(s)$ can be obtained with the ratio of the optimal open-loop frequency responses by the nominal plant defined by

$$C_F(j\omega) = \frac{\beta_{nom}(j\omega)}{G_{nom}(j\omega)} \quad (35)$$

From this frequency response, a frequency domain system identification method can be used to obtain the coefficients of an integer transfer function $C(s)$ whose frequency response fits the frequency response of $C_F(s)$. Such a method has the real advantage of producing a controller whose order is rather low (classically less than 6) whatever the problem.

To conclude, as the over-estimation of plant perturbation is avoided, the CRONE control methodology permits a non-conservative robust control system, and thus, in the majority of applications, better performance.

5.3. Design of a CRONE Controller for Fast Charging

The frequency responses of the linear models obtained through the linearization of model (24) (see Section 4.2) for various operating conditions (temperature, aging, SOC) are shown on Figure 5. This figure highlights that, in low and high frequencies respectively, the gains of the linear models are of order 1 and 0. Thus, for a low-pass filter behaviour of the controller in high-frequencies and as an integrator in low-frequencies, the open-loop orders n_l and n_h are defined such that $n_l = 2$ and $n_h = 1$. The minimization of the complementary sensitivity function resonance peak variation according to Relation (32) is acheived under the design specifications that follows:

- a sensitivity function $S(s)$ resonance peak lower than 6dB to reach a good stability degree;
- a nominal resonance peak M_{Tnom} of function $T(s)$ equal to 1.7 dB for a small overshoot of the nominal response to a step of the reference signal of J_{sr};
- a closed loop bandwidth close to 0.2 rad/s;
- a control effort sensitivity less than 10 A (I_{ch}) for a variation of J_{sr} of 10 μA in high-frequency.

The CRONE control toolbox [27,28] was used to compute the open-loop transfer function optimal parameters that meet the above specifications. The optimal parameters obtained are the following: $Y_r = 3.8$ dB, $\omega_r = 1.5$ rad·s^{-1}, $\omega_0 = 0.1$ rad·s^{-1} and $\omega_1 = 7.5$ rad·s^{-1}, leading to the following values for the fractional integration orders: $a_0 = 1.42$ and $b_0 = -0.46$. The resulting open loop transfer function Nichols chart for the nominal behaviour and the associated uncertainty domains is shown in Figure 7.

The values of ω_r, ω_0 and ω_1 appear on this figure. This figure demonstrates that the closed-loop system is stable, as all the behaviours are far from the instability point (−180°, 0 dB) area. Figure 8 shows the gain diagram of the four sensitivity functions $T(s)$, $S(s)$, $GS(s)$ and $CS(s)$ for the nominal behaviour of the plant and for two extreme behaviours of the plant. The user defined constraints (dotted lines) are also represented. As the gain of all the sensitivity functions does not exceed the constraints defined, it can be concluded that the design specifications are met. The closed loop robustness is also demonstrated by the small resonance peaks of the functions $S(s)$ and$T(s)$. An order 4 rational controller $C(s)$ is finally obtained using the method described at the end of Section 5.2. As the controller obtained contains an integrator, an anti-windup configuration is included in the control system. It cancels the charge current increase due to the output integrator saturation, thus meeting the current limits for the considered battery (a safety charge current limit of 3.5 C).

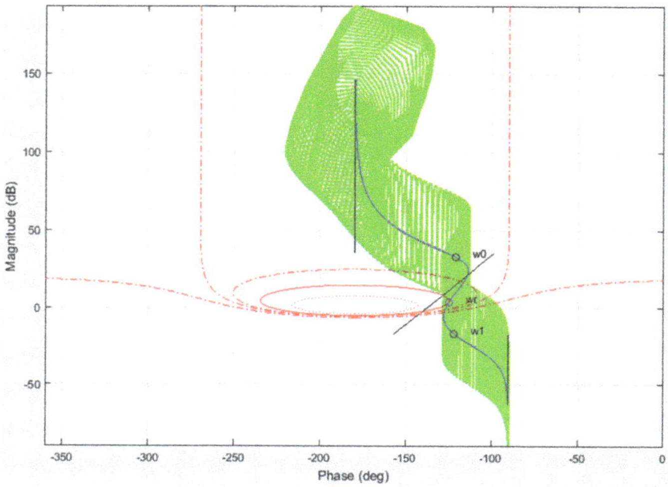

Figure 7. Nichols chart of the open loop for the uncertainty domains (**green**) associated with the nominal plant (**blue**) and Nichols abacus (**red**).

5.4. Analysis of the Control Loop Performance

The fast charging controller previously designed was applied to the model (nonlinear) of the battery described in Section 2.3. As previously mentioned, this model has a large sensitivity to aging and temperature, and the closed loop system sensitivity and robustness to these parameters must be verified. It is achieved by studying the impacts of battery parameters variations on the closed loop system response. As an example, Figure 9 shows the optimized side reaction profile tracking (profile computed as in Section 3) for the cell at BOL and for a temperature of 10 °C. The simulation is performed with a model whose parameters defining the cell aging, the initial SEI layer thickness (nm), the cell high frequency resistance (Ω), the cell capacity (Ah) and the specific surface area of the anode (m^{-1}) were changed by ~ ±20%. The J_{sr} diagram in the figure shows a tracking error close to 2 ($\mu A \cdot m^{-3}$). In spite of plant uncertainties, the controller tracking performance is thus guaranteed.

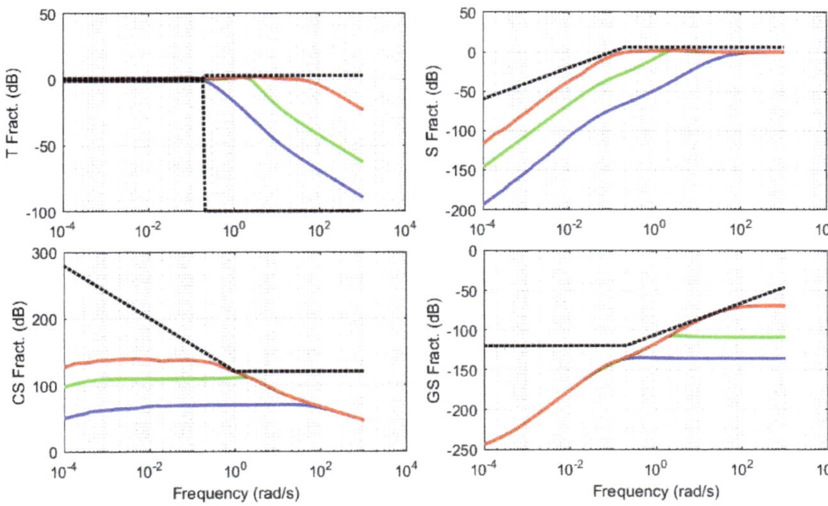

Figure 8. Gain diagram of the four sensitivity functions (solid line, green for the nominal behaviour, red and blue for two extreme behaviours of the plant) and user defined constraints (dotted lines).

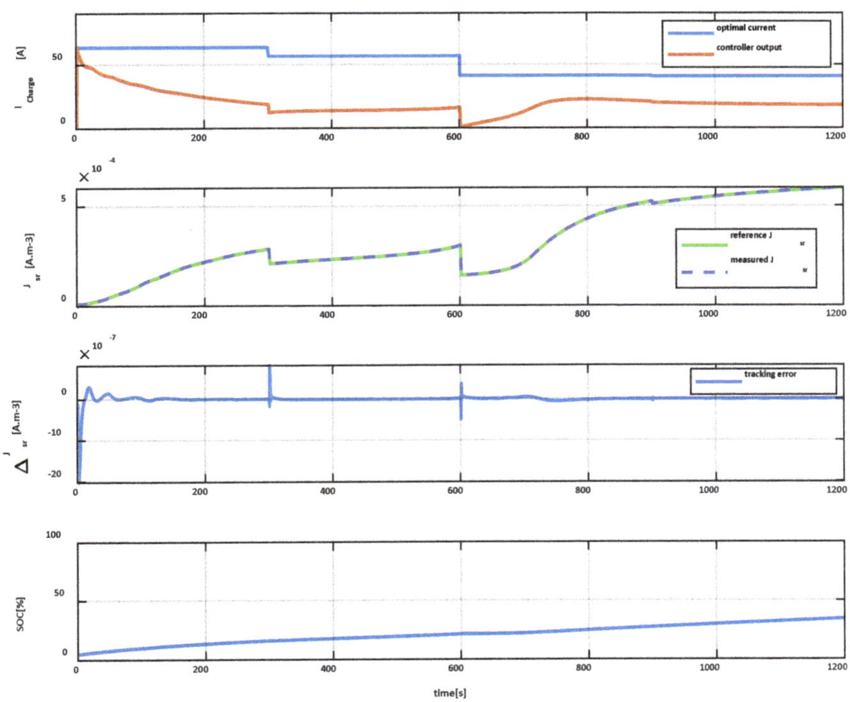

Figure 9. Impact of process variations on reference tracking.

5.5. Improvement of the Control Strategy

As it cannot be guaranteed that the final value of the SOC will be reached with the strategy previously defined, some modifications must be made, and are now proposed to correct the situation.

As shown in Figure 10, the control scheme of Figure 5 is completed by a loop that rescales the charging current I_{Charge} and side reaction current $J_{sr,traj}$ optimal trajectories against SOC.

Variations in some parameters of the battery model, such as electrolyte resistance, initial temperature and initial SEI layer, were carried out to evaluate the robustness of the modified charging strategy. The simulations performed with a cell at its BOL, and for an ambient temperature of 10 °C, are shown in Figure 11. This figure proposes a comparison between the current produced by the controller and the optimal charging current defined by the trajectory planning strategy. It highlights that the controller output moves away from the optimal current profile to maintain the optimal side reaction current tracking represented on the second diagram of Figure 10. As shown by the third diagram of Figure 11, this leads to a negligible tracking error. Using the control scheme of Figure 10, Figure 11 shows that, within the desired charging time of 20 min, the target SOC of 80% is reached.

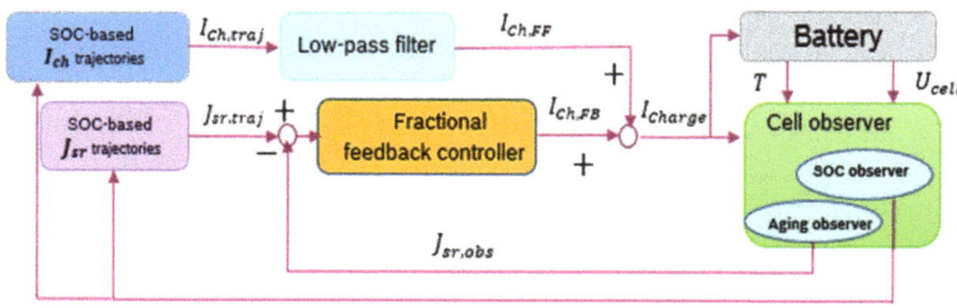

Figure 10. Modified charging controller.

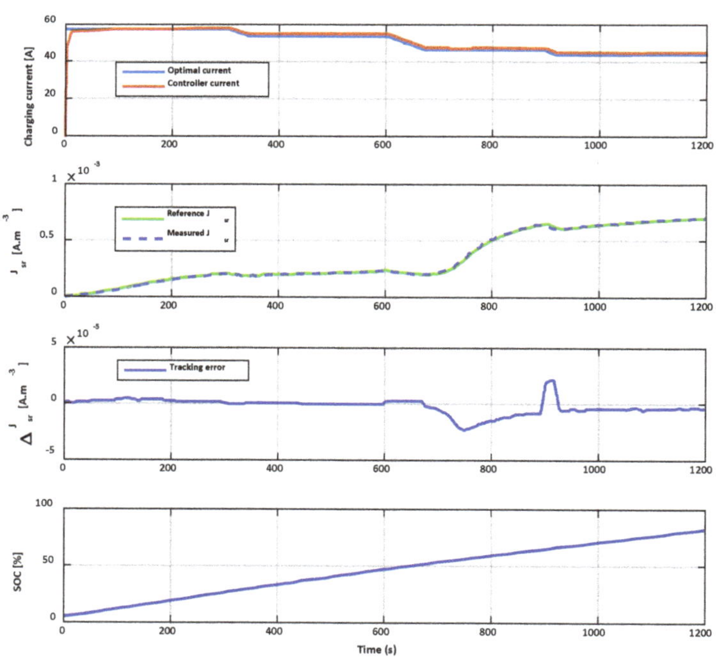

Figure 11. Validation of the modified control scheme's proper operation.

In spite of its efficiency, the control loop of Figure 10 cannot be implemented easily in a car controller. The charging controller indeed requires the computation of the side reaction current and charge current trajectories. They are obtained after the optimization of a charging profile using the method described in Section 3. However, this optimization problem involves a large number of parameters and requires the implementation of a nonlinear optimization routine. Such an approach is highly time and resource consuming, and prevents any implementation in a car BMS. This is why the authors have also designed an efficient profile optimization method, which is patent pending.

6. Conclusions

A solution for lithium-ion battery optimal fast charging was proposed in this paper. Optimality here relates to battery aging minimization, which is achieved by tracking specific charging and side reaction current reference trajectories (optimal trajectories) using a feedforward–feedback control loop. The optimal trajectories are computed using an algorithm involving an electro-thermal battery aging model recalled in this paper and developed in [6] by the same authors. This electro-thermal battery aging model also permits the display of dynamic behaviors quite different from a lithium-ion cell in relation to age and temperature. This sensitivity problem is taken into account in the charging strategy proposed using a robust CRONE feedback controller. The performance of the closed loop approach proposed is highlighted by improved side reaction current trajectories tracking, thus leading to reduced aging during fast charging. To the best knowledge of the authors, this is the first time that a closed-loop strategy, coupled with an electro-thermal battery aging model for fast charging while taking into account battery health, has been proposed in the literature.

Most of the proposed algorithms and methods in this thesis were validated by means of simulations. The evaluation of the internal battery states is not possible using real cells and measurements, but simulation provides this possibility. However, the importance of experimental validations cannot be ignored, and thus should be pursued in a future work. Upon the successful validation of the simulation results presented in this work, the electro-thermal aging battery model and the intelligent charging strategy can be implemented as a part of a BMS.

Author Contributions: S.M. contributed to the work conceptualization, investigation, development, and validation. She also contributed to paper writing (draft and review). J.S. contributed to the work conceptualization, investigation, development, supervision and validation mainly on the modelling part. He also contributed to paper writing (draft, review and editing). J.S. contributed to the work conceptualization, investigation, development, supervision and validation mainly on the control part. He also contributed to paper writing (draft and review). O.C., contributed to the problem definition, to the work supervision and to funding Acquisition. All authors have read and agreed to the published version of the manuscript.

Funding: This research received no external funding.

Conflicts of Interest: The authors declare no conflict of interest.

References

1. Botte, G.; Johnson, B.; White, R. Influence of Some Design Variables on the Thermal Behavior of a Lithium-Ion Cell. *J. Electrochem. Soc.* **2015**, *146*, 914–923. [CrossRef]
2. Sabatier, J.; Merveillaut, M.; Francisco, J.; Guillemard, F.; Porcelatto, D. Fractional models for lithium-ion batteries. In Proceedings of the European Control Conference, Zurich, Switzerland, 17–19 July 2013; pp. 3458–3463.
3. Sabatier, J.; Merveillaut, M.; Francisco, J.; Guillemard, F.; Porcelatto, D. Lithium-ion batteries modeling involving fractional differentiation. *J. Power Sources* **2014**, *262*, 36–43. [CrossRef]
4. Francisco, J.; Sabatier, J.; Lavigne, L.; Guillemard, F.; Moze, M.; Tari, M.; Merveillaut, M.; Noury, A. Lithium-ion battery state of charge estimation using a fractional battery model. In Proceedings of the IEEE International Conference on Fractional Differentiation and its Applications, Catania, Italy, 23–25 June 2014.
5. Sabatier, J.; Francisco, J.; Guillemard, F.; Lavigne, L.; Moze, M.; Merveillaut, M. Lithium-ion batteries modeling: A simple fractional differentiation based model and its associated parameters estimation method. *Signal Process.* **2015**, *107*, 290–301. [CrossRef]
6. Mohajer, S.; Sabatier, J.; Lanusse, P.; Cois, O. A Fractional-Order Electro-Thermal Aging Model for Lifetime Enhancement of Lithium-ion Batteries. In Proceedings of the MathMod Conference, Vienna, Austria, 21–23 February 2018.
7. Vo, T.T.; Chen, X.; Shen, W.; Kapoor, A. New charging strategy for lithium-ion batteries based on the integration of Taguchi method and state of charge estimation. *J. Power Sources* **2015**, *273*, 413–422. [CrossRef]

8. Abdollahi, A.; Han, X.; Avvari, G.V.; Raghunathan, N.; Balasingam, B.; Pattipati, K.R.; Bar-Shalom, Y. Optimal battery charging, Part I: Minimizing time-to-charge, energy loss, and temperature rise for OCV-resistance battery model. *J. Power Sources* **2016**, *303*, 388–398. [CrossRef]
9. Perez, H.E.; Hu, X.; Dey, S.; Moura, S.J. Optimal Charging of Li-Ion Batteries with Coupled Electro-Thermal-Aging Dynamics. *IEEE Trans. Veh. Technol.* **2017**, *66*, 7761–7770. [CrossRef]
10. Choe, S.-Y.; Li, X.; Xiao, M. Fast charging method based on estimation of ion concentrations using a reduced order of Electrochemical Thermal Model for lithium ion polymer battery. In Proceedings of the 2013 World Electric Vehicle Symposium and Exhibition (EVS27), Barcelona, Spain, 17–20 November 2013; pp. 1–11.
11. Jie Cai, J.; Zhang, H.; Jin, X. Aging-aware predictive control of PV-battery assets in buildings. *Appl. Energy* **2019**, *236*, 478–488.
12. Torchio, M.; Magni, L.; Braatz, R.D.; Raimondo, D.M. Optimal Health-aware Charging Protocol for Lithium-ion Batteries: A Fast Model Predictive Control Approach. *IFAC-PapersOnLine* **2016**, *49*, 827–832. [CrossRef]
13. Romero, A.; Goldar, A.; Garone, E. A Model Predictive Control Application for a Constrained Fast Charge of Lithium-ion Batteries. In Proceedings of the 13th International Modelica Conference, Regensburg, Germany, 4–6 March 2019.
14. Zou, C.; Manzie, C.; Nešić, D. Model Predictive Control for Lithium-Ion Battery Optimal Charging. *IEEE/ASME Trans. Mechatron.* **2018**, *23*, 947–957. [CrossRef]
15. Klein, R.; Chaturvedi, N.A.; Christensen, J.; Ahmed, J.; Findeisen, R.; Kojic, A. Optimal charging strategies in lithium-ion battery. In Proceedings of the American Control Conference (ACC), San Francisco, CA, USA, 29 June–1 July 2011.
16. Yan, J.; Xu, G.; Qian, H.; Xu, Y.; Song, Z. Model Predictive Control-Based Fast Charging for Vehicular Batteries. *Energies* **2011**, *4*, 1178–1196. [CrossRef]
17. Cheng, M.; Chen, B. Nonlinear Model Predictive Control of a Power-Split Hybrid Electric Vehicle with Consideration of Battery Aging. *ASME J. Dyn. Syst. Meas. Control* **2019**, *141*, 081008. [CrossRef]
18. Kandler, A.S.; Rahn, D.C.; Wang, C. Control oriented ID electrochemical model of lithium ion battery. *IEEE Trans. Control Syst. Technol.* **2010**, *18*, 2565–2578.
19. Smith Kandler, A. Electrochemical Modeling, Estimation and Control of Lithium-ion Batteries. Ph.D. Thesis, Pennsylvania University, Philadelphia, PA, USA, 2006.
20. Newman, J.; Thomas-Alyea, K.E. *Electrochemical Systems*, 3rd ed.; Wiley: Hoboken, NJ, USA, 2004.
21. Shabani, B.; Biju, M. Theoretical Modelling Methods for Thermal Management of Batteries. *Energies* **2015**, *8*, 10153–10177. [CrossRef]
22. Ramadass, P.; Haran, B.; Gomadam, P.; White, R.; Popov, B. Development of First Principles Capacity Fade Model for Li-Ion Cells. *J. Electrochem. Soc.* **2004**, *151*, A196. [CrossRef]
23. Randall, A.; Perkins, R.; Zhang, X.; Plett, G. Controls oriented reduced order modeling of solid-electrolyte interphase layer growth. *J. Power Sources* **2012**, *209*, 282–288. [CrossRef]
24. Guan, P.; Liu, L.; Lin, X. Simulation and Experiment on Solid Electrolyte Interphase (SEI) Morphology Evolution and Lithium-Ion Diffusion. *J. Electrochem. Soc.* **2015**, *162*, A1798–A1808. [CrossRef]
25. Ikeya, T.; Sawada, N.; Murakami, J.I.; Kobayashi, K.; Hattori, M.; Murotani, N.; Ujiie, S.; Kajiyama, K.; Nasu, H.; Narisoko, H.; et al. Multi-step constant-current charging method for an electric vehicle nickel/metal hydride battery with high-energy efficiency and long cycle life. *J. Power Sources* **2002**, *105*, 6–12. [CrossRef]
26. Sabatier, J.; Lanusse, P.; Melchior, P.; Oustaloup, A. Fractional order differentiation and robust control design. In *CRONE, H-Infinity and Motion Control, Intelligent Systems, Control and Automation: Science and Engineering*; Springer: Dordrecht, The Netherlands, 2015; Volume 77, pp. 2213–8986.
27. Lanusse, P. CRONE Control System Design, a CRONE Toolbox for Matlab. 2010. Available online: http://www.imsbordeaux.fr/CRONE/toolbox (accessed on 1 June 2017).
28. Oustaloup, A.; Melchior, P.; Lanusse, P.; Cois, O.; Dancla, F. The CRONE toolbox for Matlab. In Proceedings of the 11th IEEE International Symposium on Computer-Aided Control System Design, CACSD, Anchorage, AK, USA, 25–27 September 2000; pp. 190–195.

applied sciences

MDPI

Article

Weather-Based Prediction Strategy inside the Proactive Historian with Application in Wastewater Treatment Plants

Andrei Nicolae, Adrian Korodi * and Ioan Silea

Department of Automation and Applied Informatics, Faculty of Automation and Computers, University Politehnica Timișoara, 300223 Timișoara, Romania; andy_nicolae@yahoo.com (A.N.); ioan.silea@upt.ro (I.S.)

* Correspondence: adrian.korodi@upt.ro

Received: 12 April 2020; Accepted: 23 April 2020; Published: 26 April 2020

Abstract: The current landscape in the water industry is dominated by legacy technical systems that are inefficient and unoptimized. In recent years, sustained efforts could be identified, especially under the guidance of the Industrial Internet of Things (IIoT) paradigm, in order to develop an increased level of both connectivity and intelligence in the functioning of industrial processes. This led to the emergence of the data accumulation concept, materialized in the practical sphere by Historian applications. Although various classic Historian solutions are available, the capability to optimize and influence the monitored system in a proactive way, resulting in increased efficiency, cost reduction, or quality indicators improvements, could not be identified to date. Following a proposed software reference architecture for such a proactive Historian, a data dependency identification strategy and some obtained recipes for energy efficiency improvements in the water industry were developed. However, a complete solution for real industrial processes represents complex research. The current paper contributes to this research effort by developing part of the reference architecture that predicts the future evolution of the monitored system, based on weather dependency and forecast, thus sustaining the effort to achieve a fully functional, real-world, tested and validated proactive Historian application, with potential to bring significant direct benefits to the water industry.

Keywords: industrial internet of things; prediction strategy; proactive historian; water industry

1. Introduction

The technical solutions and processes that are specific to the water industry present a very heterogeneous characteristic, being simultaneously highly dispersed from a geographical standpoint, thus contributing to defining a landscape currently prevailed by legacy systems. Although viable solutions at the time of their implementation, the large majority of legacy systems that can be found today in the water industry are inefficient by current standards, thus transforming the water industry into the perfect environment for receiving quality, availability, and efficiency improvements.

The Industrial Internet of Things (IIoT) concept [1–4] groups a wide spectrum of industry-oriented principles [5,6], aimed at different objectives, most of which can be achieved by developing intelligent communication between different technical entities. This aspect places the superior and enhanced connectivity, interoperability, and information exchange at the core of the IIoT paradigm, which transposes from a practical standpoint to a continuous effort to connect industrial computers, sensors, and actuators to the internet [7]. The improved communication as a result of the IIoT implementation opens the way for developing more intelligent and sophisticated future systems, capable of improving the working characteristics of technical systems (such as availability, cost reduction, safety, productivity), by applying autonomous optimization techniques to the systems [8].

The importance of developing the interoperability between industrial systems is emphasized in [9], where the authors propose a communication framework that brings both efficiency and interoperability improvements for service-based architecture. Their work prepares the appropriate context for implementing future distributed monitoring and control applications. Such advances in industrial communications have recently led to the emergence of the ideas around bringing both fog computing and Big Data concepts closer to the industry, in accordance with the IIoT guidelines, because of the data accumulation phenomenon that occurs. In this direction, Aazam et al. conclude in [10] that a middleware support is required between the industrial environment and the cloud services. This middleware can be represented by the fog computing [11], which is capable of providing local computing support. On the other hand, the large amount of generated data requires Big Data technologies in the industry, and the paper from [12] presents a survey of the Big Data solutions that are already integrated into the industry, while [13] focuses on identifying the high performing Big Data techniques that can be applied to industrial data.

Despite classic Historian applications [14–16] that store the large amounts of generated data into industrial environments dominated by the IIoT principles are widely available and do not represent a novelty anymore, the most recent trends in this area are to develop solutions that can make use of the already-gathered data. The future Historian application will be intelligent and proactive, using the stored data to identify dependencies and relations between the characteristics of the monitored system, which, in turn, will be used to predict the future evolution of the technical system, as a starting point for optimizing the monitored system, in an autonomous, non-invasive, unassisted-by-human manner, being placed in the fog computing area close to the monitored system. Although such an Historian application cannot be identified to date, progress is being made toward this development direction. The study from [17] proposes efficiency improvements to a classic Historian application, while a distributed and configurable data analytics infrastructure is presented in [18]. Another distributed, wireless monitoring system is presented in [19], the problem of predictive maintenance is tackled in [20], and lightweight, Raspberry Pi-based, classic Historian solutions are presented in both [21] and [22]. In a similar research direction as that followed in this paper, Salvador et al. in [23] aim toward a proactive Historian by implementing a predictive control strategy of a water distribution network that is based on Historian gathered data. Although several researches are focusing on this gathered data capitalization toward industrial systems optimization, a fully functional, real-world, tested and validated proactive Historian solution has not been achieved to date.

In order to implement a proactive Historian solution, a software reference architecture is primarily needed and, although [24] presents general reference architectures for IIoT and [25] suggests design patterns that should be applied in IIoT, the most suitable software architecture option for a proactive Historian is described in [26]. The architecture proposed by the authors is specifically targeted for a proactive Historian type of application, dividing the required software modules into three distinct layers. The architecture is based on an already-existing classic Historian solution, which is capable of storing chosen working parameters of a technical system and structures the required modules that need to be added in order to transform a classic solution into a proactive solution. The first layer of the proposed architecture is responsible for identifying relations and dependencies between the measured characteristics by using historical stored data and external context data. The identified relations and dependencies, alongside future context data, are used as input for the second layer of the reference architecture, which must be capable of predicting the future evolution of the technical system. The last layer of the reference architecture capitalizes on the results of the previous layer, making use of the future prediction in order to decide how to influence this predicted future evolution so that chosen objectives (such as cost reduction, efficiency improvements, substances consumption reduction, maintenance improvements) are met. Ultimately, the third layer of the reference architecture should be capable of directly influencing the monitored technical system, through actuators, transferring the optimizations to the real-world system. Implementing the entire reference architecture would result in obtaining a closed-loop Historian system, which monitors a technical system, adjusts it for optimization, monitors the system reaction to the adjustment, and so on. Besides presenting the aforementioned reference architecture, the authors also implemented and tested the first level of the architecture in [26], illustrating test cases from the water industry, with reference to a drinking water treatment plant.

The current paper picks up the state of the research that was reached in [26] and takes it further down the long path to implement the entire reference architecture and obtain the first proactive Historian application capable of optimizing the monitored system without any human assistance. Achieving the final goal of this research direction demands a considerable research, implementation, and testing effort, which must be divided into several stages. The initial step was taken in [22], where the authors developed a classic Historian solution, but in a lightweight and low-cost implementation, perfectly suited for the industrial requirements. Then, important steps toward the proactive direction were made in [26], where a reference architecture was proposed, the first level of it also being implemented and tested in the water industry. The current paper falls in line with this research, representing the next significant step toward the ultimate goal. The contributions of the current paper are as follows: Improve the implementation of the first level of the reference architecture described in [26]; integrate historical weather data into the relations and dependencies identification algorithm developed in [26]; implement the second level of the reference architecture (which predicts the future evolution of the monitored system), based on the weather forecast and the results of the first layer. The unavoidable cybersecurity concerns that naturally arise from developing this type of software application are considered by the authors as being beyond the scope of the current paper.

In order to test the contributions of the current paper, several test cases are considered from the water industry, regarding a real wastewater treatment plant. This type of plant was chosen because of the typical close dependencies that exist between wastewater treatment plants and weather characteristics.

The following section presents the typical processes that take place inside a wastewater treatment plant, discloses defining problems, and emphasizes the typical weather influence over such plants, as well as detailing the main contributions of the current paper, regarding both the improvements that were made to the first level of the reference architecture and the implementation of the second level of algorithms. Section 3 illustrates the test cases that were used for validating the implementation described in the previous section in a real wastewater treatment plant environment. Section 4 dissects and discusses the results of the test cases presented in the previous section, while Section 5 concludes the paper.

2. Materials and Methods

2.1. Wastewater Treatment Plant Typical Processes

The representative processes that usually take place inside a wastewater treatment plant (WWTP) are summarized in Figure 1, and further detailed in this section. The treatment process is divided, from a logical standpoint, into multiple stages: Pretreatment, primary treatment, secondary treatment, tertiary treatment, and sludge treatment.

The wastewater enters the treatment plant from the wastewater network, where the source of the water is residential, institutional, commercial, industrial, rain, or a mix of the aforementioned. After the water enters the treatment plant, the pretreatment processes are initiated. Firstly, odor treatment can be applied so that the plant surrounding areas are protected from the foul smell that naturally accompanies wastewater. The odor treatment process may not be necessary at some plants. There are two distinct methods of odor treatment: Air treatment and liquid treatment. If the air treatment method is applied, the wastewater is contained in large tanks, which are hermetically covered with specially designed odor control covers. The air trapped under the cover, inside the tanks, is extracted by a ventilation system and undergoes treatment before it is released to the environment. Regarding the liquid treatment method, different chemicals that neutralize the foul smell-producing elements are introduced in the wastewater. The second process that takes place during pretreatment is screening, where the wastewater is passed through filters in order to remove both grit and large objects such as bottles, plastics, tree branches, sanitary items, and cotton buds. It is very important to remove such objects early in the process because they can damage the plant equipment if present in future steps of the treatment process. The removed objects are either incinerated or disposed in landfills.

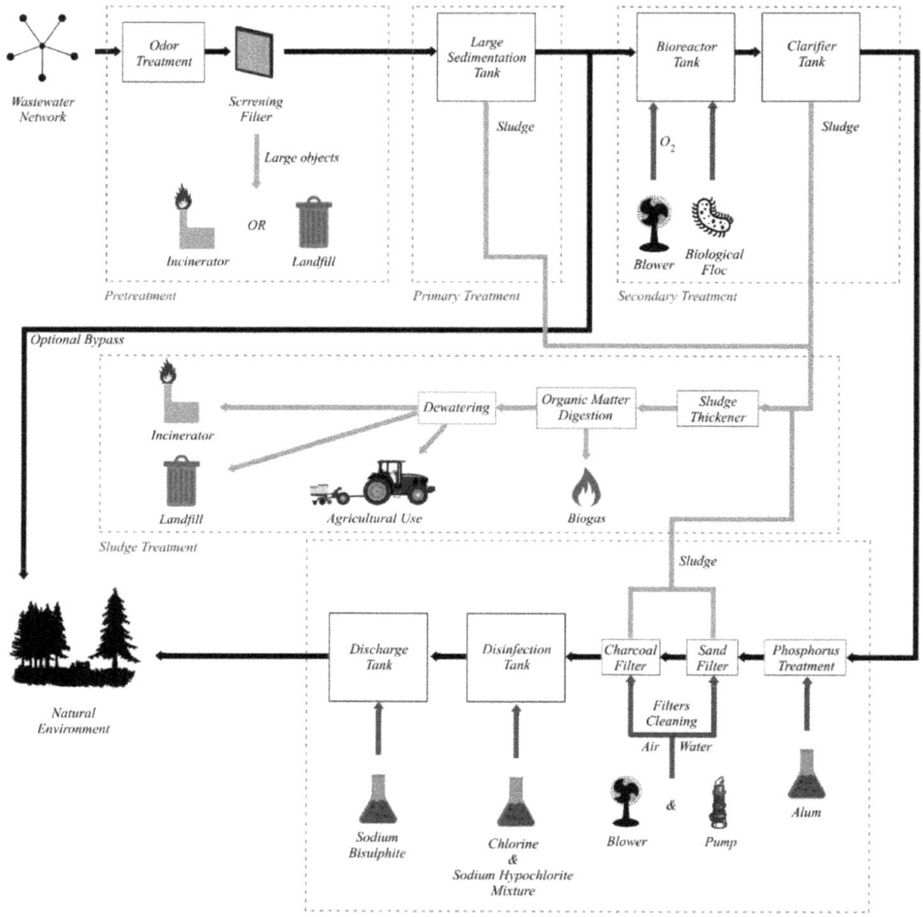

Figure 1. Typical wastewater treatment plant (WWTP) processes.

After the pretreatment is completed, the wastewater enters the primary treatment stage, where the remaining solid matter is separated from the wastewater. In this stage, large sedimentation tanks are used, in which the wastewater clarifies. The sludge settles at the bottom of the tank, while grease and oil rise to the surface. The sludge is removed and directed to the sludge treatment process, while the grease and oil can be used for soap making.

After the primary treatment, an optional bypass exists so that the treated water can be sent directly to the natural environment without entering the secondary and tertiary treatment stages. This bypass is used during heavy rainfall by plants that receive wastewater from a combined sewer system. In this case, the secondary and tertiary treatment stages are bypassed in order to protect them from hydraulic overloading, the mixture of sewage and rainwater receiving only primary treatment before being sent back to the natural environment. In some plants, the bypass is implemented directly at the inlet, so the wastewater is not even screened. In addition, for larger plants without a complete gravitational bypass, high-energy consumer pumps are used to transfer the untreated water directly into the emissary when the large amounts of wastewater exceed the capacity of the plant.

The secondary treatment stage objective is to remove the biological matter from the wastewater, by using a bioreactor tank where both oxygen (introduced with blowers) and a biological floc (bacteria and microorganisms that consume the remaining organic matter) are inserted into the wastewater. Before exiting the secondary treatment, wastewater is sent into a clarifier tank, where large particles settle down at the bottom (sludge) and are extracted for the sludge treatment process. In order to maintain the optimal process parameters during the biological treatment, a pH adjustment must take place, involving different chemicals ($Ca(OH)_2$, $CaCO_3$, Na_2CO_3, NaOH, etc.).

The last stage is the tertiary treatment, which is similar to the drinking water treatment process, the resulting water quality being close to drinking water quality. Firstly, a chemical compound (usually alum, $Al_2(SO_4)_3$, but polyaluminum chloride or ferric chloride, $FeCl_3$, can also be used) is injected into the wastewater in order to remove the phosphorus. In some plants, the phosphorus removal can be implemented during different stages, such as the large sedimentation tank, during biological treatment, or later, in the clarifier tank. If the phosphorus is removed by the biological treatment, the chemical treatment becomes an auxiliary method. Then, the water passes through a sand filter and a charcoal filter before entering a disinfection tank, where a mixture of chlorine and sodium hypochlorite is added. Lastly, the water is sent into the discharge tank where sodium bisulfite is used in order to chemically dechlorinate the water because residual chlorine is toxic to aquatic species. The water exiting the tertiary treatment is released into the natural environment in rivers, lakes, or other local waterways. Another important periodic process that takes place during tertiary treatment is the filters cleaning, in which the sand and charcoal filters are washed with air and water, the resulting sludge being sent to the sludge treatment process.

Each of the primary, secondary, and tertiary treatment stages produce sludge, which is also processed inside the WWTP, during the sludge treatment process. Firstly, the sludge enters the thickening procedure, which is conducted inside a sludge thickener, an equipment resembling a clarifier tank with an added stirring mechanism. Then, the sludge goes through the organic matter digestion process, which reduces the amount of organic matter in the sludge. Three different digestion options can be used: Aerobic digestion, anaerobic digestion, and composting. The digestion process can produce biogas (a mixture of CO_2 and methane), which can be used at the plant for powering equipment. The last sludge treatment process is dewatering, in which the sludge is commonly placed in drying beds. The dried sludge is either burned in incinerators, sent to landfills or used as fertilizer in agriculture.

2.2. Wastewater Treatment Plant Defining Problems and Weather Dependency

Some of the defining problems that can be identified in a WWTP are:

- Overloading of the plant: This can cause overheating of the blowers, which, in turn, causes a low-oxygen level in the bioreactor tank, thus reducing the efficiency of the secondary treatment stage. Plant overloading can also lead to sludge leakage from the settling tank.
- High substances consumption: For instance, the odor treatment process requires continuous adjustment of the used substances, depending on the input wastewater concentration and content. The wastewater content is highly dependent on the weather conditions.
- High energy costs: Around 30% of the annual WWTP operation costs is represented by the electricity consumption. Considering a developed country, an estimate of about 2–3% of the entire nation's electrical power is consumed for wastewater treatment. This can be significantly improved by optimizing the biological treatment processes.
- Equipment and/or algorithmic faults that can lead to various problems.

- Undersized treatment plants: Most plants were developed 10–20 years ago, becoming undersized for the current loads since then, leading to the choice of increasing the load and costs in order to maintain a thorough cleaning process or discharging the partially treated wastewater to the environment and keeping the costs lower.

A WWTP's operation is influenced by the weather conditions, the most significant being the precipitation amount. In case of very heavy rainfall, the WWTP may use the bypass channel, thus resulting in a high increase in operational costs (electricity and substances usage) or pollution. Even if the rain amount does not produce wastewater that can be legally sent to the environment after just primary treatment, the amount of rain highly influences the content and concentration of the wastewater present in the WWTP, which, in turn, influences the biological treatment that is applied in the secondary treatment stage and the chemical addition amounts and concentrations in the tertiary treatment. Besides rain, the temperature can also influence the WWTP, primarily from the odor treatment process standpoint, but also from the biological treatment in the secondary stage and sludge dewatering (when outdoor drying beds are used) processes standpoints. In addition, a storm or strong winds, particularly in the autumn, can generate large quantities of tree branches and leaves that can clog the screening filter during pretreatment.

Considering the typical processes that take place inside a WWTP, the defining problems and the usual weather influence on WWTPs that were previously presented, the conclusion emerges that a WWTP represents the perfect environment that can benefit from a solution capable of identifying the exact dependencies and relations that exist between the measured characteristics of a WWTP and meteorological characteristics. Furthermore, using those relations and dependencies for predicting the future evolution of the plant characteristics can provide a valuable foundation for optimizing the WWTP in order to reduce costs, lower energy consumption, decrease substance consumption, and improve maintenance. Due to these considerations, although the implemented solution presented in the following section represents a generic approach, it was deployed for testing and validating purposes in a WWTP environment, where the potential to make an impact is considerable.

2.3. The Implemented Solution

As previously mentioned, the solution that was implemented in this paper is based on the state of the research that was achieved in [26], making use of the already-implemented both Historian application and the first level of the proactive Historian reference architecture. This already-available technological state was tested and validated in the water industry as well (results from the already-implemented first-level algorithms were used in [27] to successfully achieve energy consumption reduction in a drinking water treatment plant, by 9% for short-term tests and by 30% for long-term tests using only part of the proposed algorithm), so it represents a reliable platform on which to build and develop, following the ultimate goal of accomplishing a fully functional proactive Historian application.

In order to merge the implementation of the second level of the reference architecture into the constantly developing application, several improvements and changes were required to the solution already implemented in [26].

First, after long-term tests, a small improvement was made to the first-level algorithm accuracy, which, in some cases, resulted in impacting differences. The computations were adjusted in order to achieve more accuracy (in the form of decimals) in relating identified data dependencies.

Another change was made regarding the choice of the reference tag. In [26], the implemented algorithm required the user to choose a reference from the available tags, and the remaining tags were analyzed regarding the chosen tag. In order to materialize a broader understanding of all the relations and dependencies that exist inside the monitored system, the implementation presented in this paper removed the need for choosing a reference tag. Instead of this approach, each of the monitored tags are set as a reference, one at a time, and the relations identifying algorithm developed in [26] is run once for each reference.

By adopting the aforementioned change related to the reference choice, new data structures are required for storing the results generated by the first-level algorithm. The relations identifying algorithm was adjusted in order to build an oriented graph of dependencies, where the results of the algorithm execution are stored. The built graph is oriented, weighted, and can contain cycles. The results are modeled using the following convention:

- An arc from node i to node j with weight -N signifies when node i was set as the reference,
- A dependency of node j on node i was identified,
- The measured values of node j evolving inversely proportional (minus sign) to the node i values, in a quantitative proportion of N% (this percent represents the quantitative result identified by the analysis; more details regarding this percent is available in [26]).

In the current implementation, the dependencies graph is stored using the adjacency matrix. Inside the matrix, line i contains the dependency values (please refer to [26] for details) of all nodes j on the columns when node i is used as a reference. The dependencies graph represents essential input for the second level of the reference architecture.

The last improvement that was brought to the solution developed in [26] consists of the possibility to involve the meteorological characteristics in the relations analysis. Historical weather data can be used in the relations and dependencies identification algorithm (at the first level of the reference architecture) at user demand. The historical weather data source is the DarkSky online service [28]. If the user chooses to use historical weather data, the values of the relevant weather characteristics for the water industry (maximum temperature, minimum temperature, precipitation amount, humidity, atmospheric pressure, wind speed, and ultraviolet (UV) index) are obtained from [28] for each of the days in which tags values that are involved in the analysis exist. As a requirement for obtaining the weather data, the geographical location of the monitored technical system must be provided by the user. The longitude and latitude of the location that are required for calling the weather application programming interface (API) are obtained from the user-provided address, using [29]. After having at disposal the values of each of the 7 meteorological characteristics considered of interest, each of them is used as a reference, one at a time, in the relations identifying algorithm, which computes only the dependency of the technical system tags on the weather features. The weather features dependency on the technical system tags does not make sense, so it is not computed. As a consequence, if the user chooses to include historical weather data into the first-level algorithm analysis, the adjacency matrix of the dependencies graph will be deformed, containing i + o lines and i columns (where i signifies the number of tags from inside the monitored system and o signifies the number of tags from outside the system, essentially the number of meteorological features), meaning that the graph will not contain any arcs from a technical system-monitored tag to a weather feature.

The improvements described above have led to the modification of the Historian application graphical user interface (GUI), where the new interface elements are presented in Figure 2. The user may enable the Historian to be augmented with the predictive algorithm from the second layer of the reference architecture. This action is only allowed if the historical weather data are used in the relations and dependencies identification algorithm, at the first level.

Improvements in the developed solution from [26] were necessary for the second level of the reference architecture. The section follows with the predictive algorithm development, placed at the second level in the reference architecture. The algorithm predicts the future evolution of the monitored technical system, based on the weather forecast and the relations and dependencies identified by the first-level algorithms.

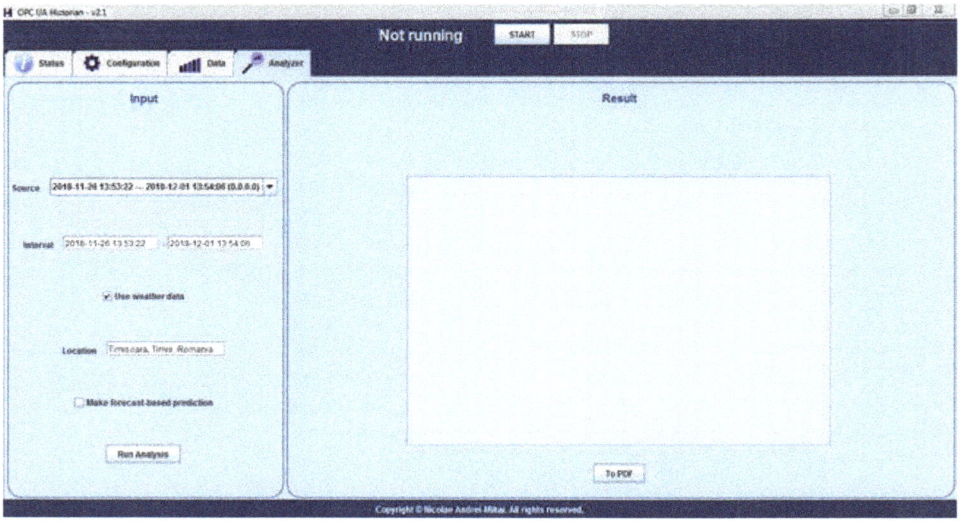

Figure 2. The new graphical user interface (GUI) elements of the Historian application.

The execution of the developed prediction algorithm is conditioned by the following prerequisites:

- the dependencies graph generated by the first-level algorithms must be available;
- weather forecast data must be obtained from [28];
- the most recent values of the monitored tags (which are used as initial values in the prediction process) must be extracted from the database (it is not necessary that they represent the current values; if the current values are not available, then the most recent ones are used);

Considering the valid prerequisites, the prediction algorithm is launched in execution, and Figure 3 below summarizes an overview of the implemented algorithm.

```
for z in forecasted_days
    set_initial_values_for_current_day ()
    for m in weather_features
        bfs_traversal_of_dependencies_graph (starting_root = m)

function process_node ()
    prev_perc = compute_change_percent_of_coming_from_node_regarding_previous_day ()
    prev_sign = compute_value_increase_or_decrease_coming_from_node_regarding_previous_day ()
    dep = get_graph_dependency_for_coming_from_node_to_current_node ()
    val_c = get_current_value_of_current_node ()
    proc_c = compute_percent_of_change_for_current_node (prev_perc, dep)
    val_ch = compute_value_in_units_of_change_for_current_node (proc_c, val_c)
    val_new = compute_new_value_of_current_node (val_ch, val_c, dep, prev_sign)
    set_new_value_for_current_node (val_new)
```

Figure 3. Summarized overview of the implemented prediction algorithm.

When the predictive algorithm starts predicting values for a new day, it first initializes the monitored tag values from the technical system. For the first day of prediction, the initial values for the system tags are the most recent values from the database. For the remaining predicted days, the initial values are the same as those computed/predicted by the algorithm for the previous day.

The forecast weather data can be obtained from [28] for a maximum of 7 days (d) ahead, thus leading to a breadth-first traversal of the dependencies graph for each of the chosen weather features in each of the 7 d ahead. The standard breadth-first traversal algorithm was adjusted in order to remember the node from which the traversal arrived at the current node, information required by the function that processes the current node (*process_node* function is presented in Figure 3). The root node of the breadth-first traversals is always a weather feature, meaning that, starting from the weather features-predicted evolution, the algorithm can predict the evolution of the technical system tags, based on the identified dependency between the weather features and the system characteristics that were identified by the first-level algorithm. By executing the breadth-first traversal, all the existing relations between the technical system tags are considered in a global manner. For instance, if technical system characteristic A is directly influenced by temperature and technical system characteristic B is inversely influenced by A, the weather forecast indicating an increase in temperature for the following day, only computing characteristic A's new value based on the temperature increase, does not offer an accurate prediction of the system's overall evolution, because the increase in A causes a decrease in B as well. These kinds of cases are fully covered by the implemented algorithm, thus seeking to obtain a realistic prediction.

The development of the *process_node* function is based on:

- computing the percent of change in the previous node (the node from which the arc leading to the current node starts), by comparing the previous node's current value and the node's previous day value.
- computing the sign of change (if the previous node's value increased or decreased from the previous day).
- using the percent of change in the previous node and the dependency from the graph, the predictive algorithm computes the percent of change for the current node (more details regarding the value of dependency in the graph can be found in [26]).
- the percent of change for the current node is further used alongside the current value of the current node in order to identify the value of change (in units) for the current node. The value of change is onwards used together with the current value of the current node, the corresponding dependency from the graph, and the sign of change for the previous node, in order to compute the new value of the current node.

The output of the implemented algorithm represents the predicted values for the monitored tags from the technical system for 7 d of the prediction.

Thus far, the presented research implementation of the second layer follows a generic approach and can extend the water domain to any industry that relates to weather data. However, in order to apply and test the algorithm, a specific process is absolutely necessary.

After implementing the second layer of the reference architecture, it became obvious that the relations and dependencies identification algorithm and the predictive algorithm could both receive a significant accuracy improvement by capitalizing on different process-specific information, which can be used during algorithm execution (e.g., knowing that a specific Open Platform Communications Unified Architecture (OPC UA) tag signifies the 'fault' code for a pump can be used during analysis in order to avoid false-positives identification for dependencies between the pump energy consumption and other values; knowing that from a process point of view, the aeration process requires a blower that implicitly consumes energy; knowing process flow of the specific process; etc.). The process-aware Historian concept is essential for correct predictions, recipes, relevant data dependency analysis, and constraint and objective function interpretation. To be able to gather and use those process-specific information, the Historian application received a newly developed software module (Process Editor), which allows the creation of a model of a monitored process, inside the Historian application. Multiple processes can be defined, from which only one can be set as the currently used one, thus facilitating an easy switch between different monitored processes. Therefore, the process-aware Historian will request essential data mapping from the user according to the predefined process components characteristics. A defined process inside the Process Editor contains steps, which, in turn, contains items. There are multiple predefined item types from which the user can choose (water source, air blower, pump, flowmeter, water tank, etc.), each item having its own predefined characteristics (example for a biological basin: Level, set point, oxygen level, NH_4 level). For each characteristic, the user can set an OPC UA tag from the server list, thus assigning a meaning to each monitored OPC UA tag. Figure 4 presents the editing of an item from the Process Editor, while Figure 5 presents a process of a WWTP, as defined in the Historian Process Editor. Furthermore, the possibility of adding different constraints to the process was also implemented and is illustrated in Figure 6.

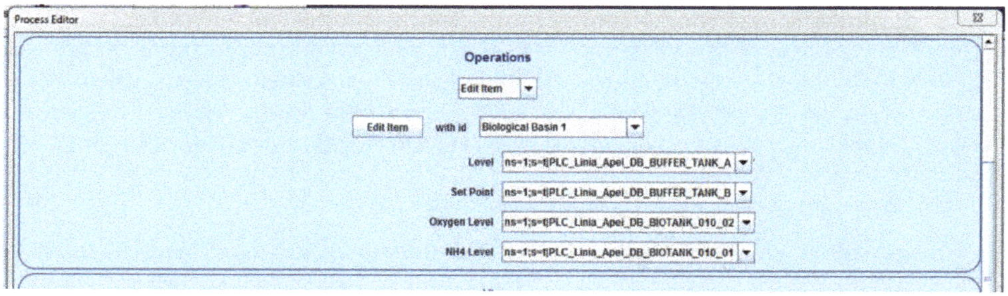

Figure 4. Editing an item in the Historian Process Editor.

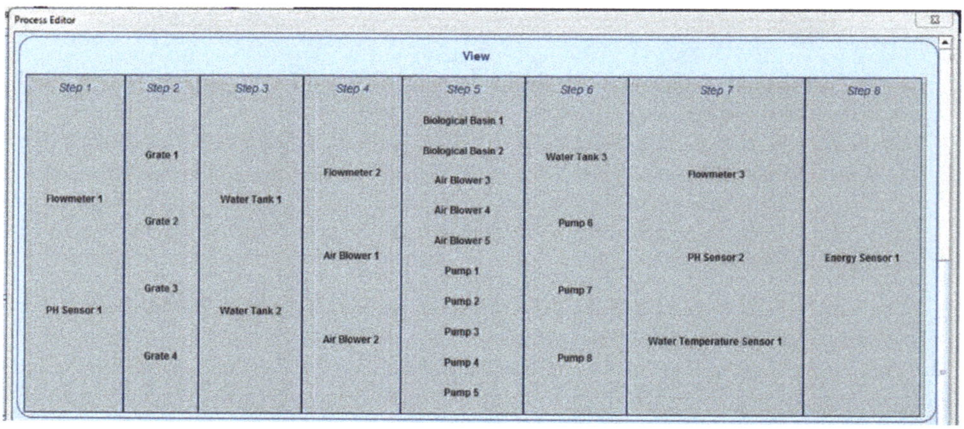

Figure 5. A process of a WWTP, as defined in the Historian Process Editor.

Considering the generic approach of the Historian application, which is intended at not restricting it only to the water industry, even though predefined item types and characteristics are offered, the implementation created a generic framework, which makes use of two Java String arrays (containing all the predefined elements), for both building the required GUI elements and interacting with the Extensible Markup Language (XML) structure used for storing the process definition. This approach implies that adding and/or removing items or item characteristics requires just a small change in an array of Strings (no additional changes are required at GUI elements building or XML interaction), thus keeping the application suitable for easy expansion toward other industries.

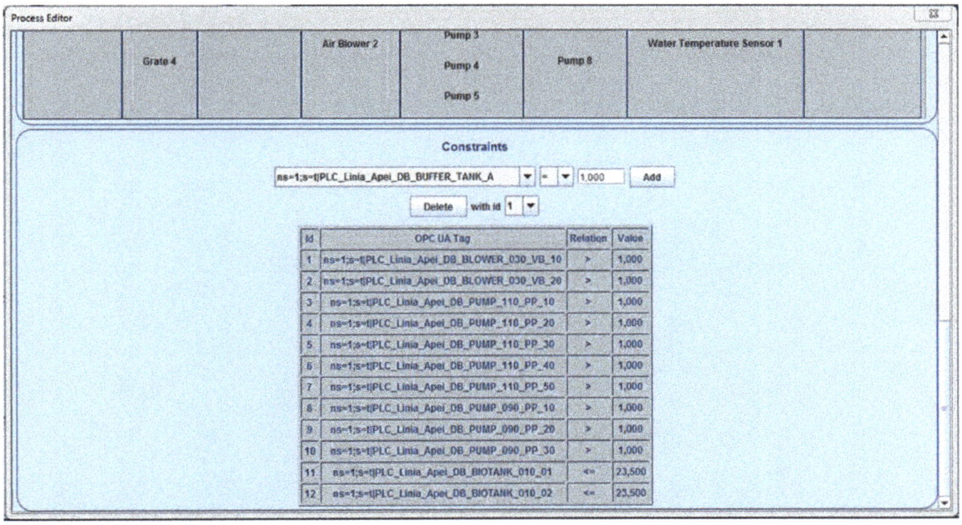

Figure 6. WWTP process constraints in the Historian Process Editor.

3. Results

With the purpose of testing and validating the newly implemented features described in the previous section of this paper, multiple test cases were considered in the water industry, regarding a real WWTP, owned and operated by the local water company.

The tests were conducted by making use of the most recent Historian application version available, which included the implementation of both level 1 and 2 of the reference architecture, while being installed on the Raspberry Pi platform.

Figure 7 details the seven test cases that were considered for validating the second level of the reference architecture implementation. Each test case is covered over a period of time where data are collected (a set of relevant process data from a WWTP can contain 200–2000 tags, although the total number of tags from a WWTP can exceed 6000). From the entire set of tags, the monitored tags are defined as relevant to predict regarding the weather evolution. The number of analyzed tags for value prediction was 26 in all seven test cases. The 26 tags were chosen by the authors as being the most relevant and having the most potential of being used in any future optimization, representing mostly electric current consumptions for different pumps and air blowers and different water volumes inside the treatment plant and water quality indicators, such as turbidity.

Initially, the Historian application was used to store the measured values of the chosen tags for the time period presented in Figure 7, a task that did not involve any proactive features and could have been carried out by any classic Historian application. Afterward, the proactive features were assessed by executing the predictive algorithm for each test case, using the GUI controls illustrated in Figure 2 above. The inputs for the test execution included historical weather data alongside the stored values of the tags from the monitored system and started with the execution of the first level algorithm (the improved version described in the previous section, deriving from that implemented in [26]), which generated as output the dependencies graphs (a different one for each test case), such as the example depicted in Figure 8 (corresponding to test case 7). In building the representation from Figure 8, only the tags (and their corresponding relations) that were found dependent on at least one weather characteristic were considered, where a drawing approach containing all the monitored system tags (and their corresponding relations) made it difficult to follow.

#	Period of collecting the values of the analyzed tags	Number of analyzed tags	Number of tags identified as being related to weather
Test case 1	28.01.2020 - 26.02.2020	26	2
Test case 2	28.01.2020 - 26.02.2020	26	2
Test case 3	28.01.2020 - 03.02.2020	26	12
Test case 4	28.01.2020 - 06.02.2020	26	9
Test case 5	28.01.2020 - 10.02.2020	26	10
Test case 6	22.02.2020 - 26.02.2020	26	17
Test case 7	17.02.2020 - 26.02.2020	26	6

Figure 7. The 7 test cases considered for validation of the implementation.

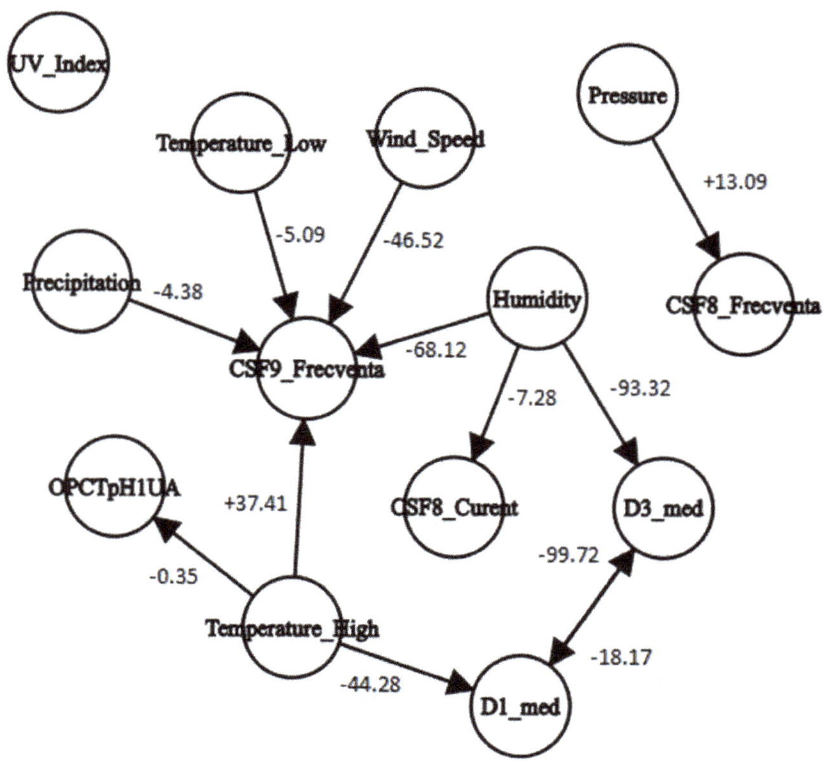

Figure 8. The dependencies graph for test case 7.

The dependencies graphs (such as that shown in Figure 8) were used as input, in addition to the weather forecast (obtained from [28]) and the latest tags values at the moment of analysis, for the predictive algorithm described in the previous section. The outcome consisted of the algorithm successfully computing the predicted values of each of the 26 tags from the monitored system for the following 7 d starting from the execution date, in each of the seven test cases. Basically, the developed prediction algorithm accurately estimated the impact the forecasted weather will have on the technical system, by identifying and understanding (from the historical stored data) the relations that exist between the weather and the technical system tags. An example of the numerical results of the algorithm execution (for test case 5) is presented in Figure 9, where the illustrated pdf document has been exported directly from the Historian application.

```
            Prediction results

            Date: Wed Mar 04 00:00:00 EET 2020

F1_Debit_T: 65850.67
F1_Curent: 3.65
F1_Turatia_Pompa: 0.0
F2_Debit_T: 199605.8
F2_Curent: 0.08
F2_Turatia_Pompa: 0.0
F3_Debit_T: 84777.09999997432
F3_Curent: 0.1
F3_Turatia_Pompa: 0.0
D1_med: 0.0
D1_index: 0.0
OPCTTP1UA: 14.269999999997724
OPCTpH1UA: 8.4
CSF7_Curent: 4.37
CSF7_Frecventa: 0.0
CSF8_Curent: 4.175208435508391
CSF8_Frecventa: 0.0
CSF9_Curent: 163.34
CSF9_Frecventa: 0.0
SPI1_ENERGIE: 3.637763609612555
CSF10_Curent: 0.0
CSF10_Frecventa: 39.78
SPI2_ENERGIE: 1.0
D3_med: 25.34
D3_index: 0.0
TGD1_Ep: 700386.95

            Date: Thu Mar 05 00:00:00 EET 2020

F1_Debit_T: 65807.39
F1_Curent: 3.94
F1_Turatia_Pompa: 0.0
F2_Debit_T: 199634.42
F2_Curent: 0.11
F2_Turatia_Pompa: 0.0
F3_Debit_T: 84777.09999997432
F3_Curent: 0.15
F3_Turatia_Pompa: 0.0
D1_med: 0.0
D1_index: 0.0
OPCTTP1UA: 14.269999999997724
OPCTpH1UA: 8.41
CSF7_Curent: 4.41
CSF7_Frecventa: 0.0
CSF8_Curent: 4.175208435508391
CSF8_Frecventa: 0.0
CSF9_Curent: 86.53
CSF9_Frecventa: 0.0
SPI1_ENERGIE: 3.637763609612555
CSF10_Curent: 0.0
CSF10_Frecventa: 39.83
SPI2_ENERGIE: 1.0
D3_med: 25.37
D3_index: 0.0
TGD1_Ep: 700168.01

            Date: Fri Mar 06 00:00:00 EET 2020

F1_Debit_T: 64636.63
```

```
F1_Curent: 3.94
F1_Turatia_Pompa: 0.0
```

Figure 9. Prediction algorithm output for test case 5.

4. Discussion

After computing the prediction, the Historian application was left in operation to store the values of all the monitored tags for a period of 7 d (the prediction is made for 7 d ahead); thus, at the end of the 7 d for which predicted values were available, the actual values, recorded from the WWTP, were also available, facilitating an assessment of the prediction algorithm viability.

For supporting an identification of conclusive aspects, in each of the seven test cases considered, the predicted values were compared to the actual values; the most significant results, considering all test cases, regarding the possibility of future optimizations and improvements in the monitored technical system, as summarized in Figure 10, are displayed in the following more detailed explanation.

#		04.03.2020	05.03.2020	06.03.2020	07.03.2020	08.03.2020	09.03.2020	10.03.2020	Average Error
CSF10_Curent	Predicted	3.978	3.983	4.009	4.018	4.042	4.047	-	8.340
	Actual	4.311	4.346	4.402	4.400	4.384	4.425	-	
	Error	7.729	8.353	8.934	8.690	7.798	8.535	-	
OPCTpH1UA	Predicted	1.084	1.084	1.085	1.085	1.086	1.086	-	0.413
	Actual	1.080	1.080	1.080	1.080	1.082	1.081	-	
	Error	0.326	0.401	0.464	0.452	0.340	0.493	-	
TGD1_Ep	Predicted	700386.950	700568.010	700807.450	701174.740	701213.080	701382.850	701560.570	8.500
	Actual	763665.357	765490.546	766042.646	766599.938	766929.821	767016.004	767199.572	
	Error	8.286	8.481	8.516	8.534	8.569	8.557	8.556	
CSF7_Curent	Predicted	4.370	4.410	3.940	4.000	3.910	3.930	3.980	15.801
	Actual	4.833	4.936	4.976	4.764	4.757	4.847	4.783	
	Error	9.575	10.649	20.827	16.046	17.805	18.918	16.790	
F2_Debit_T	Predicted	199605.800	199634.420	199806.530	207457.010	208508.560	209463.310	212342.880	8.694
	Actual	224429.000	224472.474	224589.842	224808.748	225061.228	225096.466	225123.343	
	Error	11.061	11.065	11.035	7.718	7.355	6.945	5.677	

Figure 10. The most significant results of the test cases.

Figure 10 presents five OPC UA tags (CSF10_Curent, OPCTpH1UA, TGD1_Ep, CSF7_Curent, and F2_Debit_T), whose values were identified as being related to the evolution of different weather characteristics. For each of those tags, Figure 10 illustrates the predicted value (blue lines) and the actual value (green lines) in each of the seven predicted days (the seven columns with headers containing dates). The error lines (red lines) present the prediction error for each day (computed in percent, using the following formula: E = (ABS (P − A) * 100) / A, where E = error, P = predicted value, A = actual value, and ABS = function returning the absolute value). The last column ("Average Error") represents the arithmetic mean of the error values computed per day of the respective tag.

In the continuation of this section, the interpretation of the results presented in Figure 10 follows, with regard to the optimization possibilities of a WWTP.

Tag CSF10_Curent represents the intensity (in amps) of the electric current consumed by a pump, which is used for the transfer of wastewater from mechanical treatment to biological treatment inside a WWTP. This pump intensifies in usage as more wastewater enters the system and has a similar behavior as the bypass pump. Being able to accurately predict the usage of those pumps signifies the capability to predict the usage amount of the bypass system, which also has a huge impact on both the overall power consumption of the WWTP and the overall substances consumption of the WWTP. The usage of the transfer pumps and the bypass system is dependent on rainfall for the majority of the plants, leading to the conclusion that the prediction algorithm can bring significant added value to this area.

Tag OPCTpH1UA signifies the water turbidity, measured at the exit of the WWTP, a characteristic that shows the water quality after receiving the treatment inside the WWTP. This characteristic must meet legal requirements regarding its limits, and predicting it helps in better estimates of the future consumption of substances.

Tag TGD1_Ep represents the overall WWTP energy consumption, its accurate prediction being imperatively needed for any energy-related optimization.

Tag CSF7_Curent represents the intensity (in amps) of the electric current consumed by an air blower that introduces oxygen into a biological basin. Estimating this value accurately opens the possibility of understanding and predicting the energy consumption of a biological basin and its weight in the overall energy consumption.

Tag F2_Debit_T signifies the water volume at the WWTP entrance, where its correct prediction allows for an accurate estimate of the overall energy consumption, the overall substances consumption, and the bypass system usage. The water volume at the WWTP entrance is directly influenced to a large extent by meteorological characteristics (especially precipitation amount).

The differences between the considered test cases resulted from the various weather characteristics encountered during each considered period, such that the authors could identify different possible optimization paths to follow in the future third-level implementation of the reference architecture.

To conclude the results analysis, the seven test cases involved successfully capitalized on real data recorded from the WWTP, with promising results. Unfortunately, the exact overall accuracy of the predictive algorithm is very difficult to assess because it is directly influenced by the weather forecast accuracy. For example, a difference of 1 °C between the forecasted temperature and the actual temperature at a 5° value (average temperature during the period when the test cases were considered) leads to a 20% error that is introduced by the weather forecast accuracy, not by the implemented algorithm. In addition, analyzing historical data gathered in a period of time in which certain meteorological phenomena did not occur (for example, it did not rain) could lead to the situation in which the relations identifying algorithm does not find any dependency between some of the meteorological characteristics and technical system tags (even if they could exist). This implies that the algorithm's overall accuracy could be better evaluated after considering some long-term test cases (covering multiple months or even years).

Due partly to the exponential growth of the dependencies graph dimensions when a larger number of variables are involved that makes it infeasible to track by human operators and partly to the paper space constraints, a test case involving a larger number of variables (over 100) cannot be presented in this section, by illustrating each step of the analysis, because the section presents examples from different test cases that are suitable for highlighting the necessary aspects. Nevertheless, taking into account all the tests performed by the authors, the latest Historian application version can be considered validated as a solution capable of computing the impact of the weather characteristics over a technical system from the industry, thus meeting the goals set for the current paper.

5. Conclusions

The state of the art in industrial automation research is focused toward the IIoT principles, guiding the industry into a more intelligent era, characterized in the incipient phase that is taking place currently by intelligent communication, improved interoperability, and connectivity. The transition toward this new era has already begun in recent years and is currently in full swing, where the huge potential is enabled by the new technologies being recognized by the industry as well. After this initial transition, the framework and infrastructure will be in place in order to develop the next significant level of improvements, in the form of intelligent, autonomous, proactive software applications, possessing the capabilities of analyzing technical systems and optimizing them for maximized performance. In the industry, a fog-based process-aware proactive Historian concept satisfies these requirements. The current research is hesitant in offering such solutions.

Addressing the challenge of developing the future generation of proactive Historian applications has been started by the authors with encouraging results. The current paper sustains this effort and brings the solution a step closer to the final goal. The presented contributions are bringing both improvements to the first level of the reference architecture, as well as contributions to the second level of the reference architecture in the form of predicting process values regarding integrated weather data as context data. The research is applied in the water industry, particularly for wastewater treatment plants. The testing is realized using wastewater-specific test cases, and the obtained results are promising. Although the water industry represents the main target of the current development, the software solution follows a generic mindset, which does not limit it to any specific industry.

To conclude, the current paper falls in line with a series of research papers in what is already starting to outline itself as a well-defined research direction, contributing to the efforts of achieving the fully functional, tested, validated, proactive Historian solution that will prove its value in tomorrow's industry.

Author Contributions: Conceptualization, A.K.; methodology, A.K. and I.S.; software, A.N.; validation, A.N.; formal analysis, A.N.; investigation, A.K. and A.N.; writing—original draft preparation, A.N.; writing—review and editing, A.K.; supervision, I.S.; funding acquisition, A.K. and I.S. All authors have read and agreed to the published version of the manuscript.

Funding: This research received no external funding.

Conflicts of Interest: The authors declare no conflict of interest.

References

1. Yury, S.; Samoylova, E. The internet of things as socio-technological institution of civil society in post-informational era. In Proceedings of the 2nd International Conference on Computer and Communication Systems (ICCCS), Krakow, Poland, 11–14 July 2017.
2. Sisinni, E.; Saifullah, A.; Han, S.; Jennehag, U.; Gidlund, M. Industrial Internet of Things: Challenges, Opportunities, and Directions. *IEEE Trans. Ind. Inf.* **2018**, *99*, 4724–4734. [CrossRef]
3. Perera, C.; Liu, C.H.; Jayawardena, S.; Chen, M. A Survey on Internet of Things From Industrial Market Perspective. *IEEE Access* **2014**, *2*, 1660–1679. [CrossRef]
4. Nicolae, A.; Korodi, A.; Silea, I. An Overview of Industry 4.0 Development Directions in the Industrial Internet of Things Context. *Rom. J. Inf. Sci. Tech.* **2019**, *22*, 183–201.
5. Strasser, T.I.; Andrén, F.P.; Vrba, P.; Šuhada, R.; Moulis, V.; Farid, A.M.; Rohjans, S. An Overview of Trends and Developments of Internet of Things Applied to Industrial Systems. In Proceedings of the 44th Annual Conference of the IEEE Industrial Electronics Society (IECON), Washington, DC, USA, 21–23 October 2018.
6. Xu, H.; Yu, W.; Griffith, D.; Golmie, N. A Survey on Industrial Internet of Things: A Cyber-Physical Systems Perspective. *IEEE Access* **2018**, *6*, 78238–78259. [CrossRef]
7. Vandikas, K.; Tsiatsis, V. Performance evaluation of an IoT platform. In Proceedings of the IEEE Eighth International Conference on Next Generation Mobile Apps, Services and Technologies (NGMAST), Oxford, UK, 10–12 September 2014.
8. Khan, W.A.; Wisniewski, L.; Lang, D.; Jasperneite, J. Analysis of the requirements for offering industrie 4.0 applications as a cloud service. In Proceedings of the 26th IEEE International Symposium on Industrial Electronics (ISIE), Edinburgh, UK, 19–21 June 2017.
9. Eliasson, J.; Delsing, J.; Derhamy, H.; Salcic, Z.; Wang, K. Towards industrial Internet of Things: An efficient and interoperable communication framework. In Proceedings of the IEEE International Conference on Industrial Technology (ICIT), Seville, Spain, 17–19 March 2015.
10. Aazam, M.; Zeadally, S.; Harras, K.A. Deploying Fog Computing in Industrial Internet of Things and Industry 4.0. *IEEE Trans. Ind. Inf.* **2018**, *14*, 4674–5682. [CrossRef]
11. Gazis, V.; Leonardi, A.; Mathioudakis, K.; Sasloglou, K.; Kikiras, P.; Sudhaakar, R. Components of fog computing in an industrial internet of things context. In Proceedings of the 12th Annual IEEE International Conference on Sensing, Communication, and Networking - Workshops (SECON Workshops), Seattle, WA, USA, 22–25 June 2015.
12. Al-Gumaei, K.; Schuba, K.; Friesen, A.; Heymann, S.; Pieper, C.; Pethig, F.; Schriegel, S. A Survey of Internet of Things and Big Data Integrated Solutions for Industrie 4.0. In Proceedings of the 23rd International Conference on Emerging Technologies and Factory Automation (ETFA), Turin, Italy, 4–7 September 2018.
13. Geng, D.; Zhang, C.; Xia, C.; Xia, X.; Liu, Q.; Fu, X. Big Data-Based Improved Data Acquisition and Storage System for Designing Industrial Data Platform. *IEEE Access* **2019**, *7*, 44574–44582. [CrossRef]
14. Eren, H. Assessing the health of sensors using data historians. In Proceedings of the IEEE Sensors Applications Symposium (SAS), Brescia, Italy, 7–9 February 2012.
15. Mazur, D.C.; Entzminger, R.A.; Kay, J.A. Enhancing Traditional Process SCADA and Historians for Industrial and Commercial Power Systems With Energy (Via IEC 61850). In Proceedings of the IEEE/IAS 50th Industrial & Commercial Power Systems Technical Conference (I&CPS), Fort Worth, TX, USA, 20–23 May 2014.
16. Fras, A.; Dang, T. Improving industrial application's performances with an Historian. In Proceedings of the IEEE International Conference on Industrial Technology (ICIT), Hammamet, Tunisia, 8–10 December 2004.

17. Jankowski, T.; Davis, G.; Holmes, J.; Kemper, G. Increasing data historian efficiency. In Proceedings of the IEEE-IAS/PCA 53rd Cement Industry Technical Conference, St. Louis, MO, USA, 22–26 May 2011.
18. Kefalakis, N.; Roukounaki, A.; Soldatos, J. A Configurable Distributed Data Analytics Infrastructure for the Industrial Internet of things. In Proceedings of the 15th International Conference on Distributed Computing in Sensor Systems (DCOSS), Santorini Island, Greece, 29–31 May 2019.
19. Mateev, V.; Marinova, I. Distributed Internet of Things System for Wireless Monitoring of Electrical Grids. In Proceedings of the 20th International Symposium on Electrical Apparatus and Technologies (SIELA), Bourgas, Bulgaria, 3–6 June 2018.
20. Prakash, C.; Thakur, S. Smart Shut-Down and Recovery Mechanism for Industrial Machines Using Internet of Things. In Proceedings of the 8th International Conference on Cloud Computing, Data Science & Engineering (Confluence), Noida, India, 11–12 January 2018.
21. Kulkarni, P.H.; Kute, P.D.; More, V.N. IoT based data processing for automated industrial meter reader using Raspberry Pi. In Proceedings of the 2016 International Conference on Internet of Things and Applications (IOTA), Pune, India, 22–24 January 2016.
22. Nicolae, A.; Korodi, A. Node-Red and OPC UA Based Lightweight and Low-Cost Historian with Application in the Water Industry. In Proceedings of the IEEE 16th International Conference on Industrial Informatics (INDIN), Porto, Portugal, 18–20 July 2018; pp. 1012–1017.
23. Salvador, J.R.; de la Peña, D.M.; Ramirez, D.R.; Alamo, T. Historian Data Based Predictive Control of a Water Distribution Network. In Proceedings of the European Control Conference (ECC), Limassol, Cyprus, 12–15 June 2018.
24. Weyrich, M.; Ebert, C. Reference Architectures for the Internet of Things. *IEEE Softw.* **2016**, *33*, 112–116. [CrossRef]
25. Bloom, G.; Alsulami, B.; Nwafor, E.; Bertolotti, I.C. Design patterns for the industrial Internet of Things. In Proceedings of the 14th IEEE International Workshop on Factory Communication Systems (WFCS), Imperia, Italy, 13–15 June 2018.
26. Nicolae, A.; Korodi, A.; Silea, I. Identifying Data Dependencies as First Step to Obtain a Proactive Historian: Test Scenario in the Water Industry 4.0. *Water* **2019**, *11*, 1144. [CrossRef]
27. Korodi, A.; Crisan, R.; Nicolae, A.; Silea, I. Industrial Internet of Things and Fog Computing to Reduce Energy Consumption in Drinking Water Facilities. *Processes* **2020**, *8*, 282. [CrossRef]
28. Dark Sky API Weather Service. Available online: https://darksky.net/dev (accessed on 12 April 2020).
29. OpenStreetMap Nominatim Service for Converting an Address to Longitude and Latitude. Available online: https://nominatim.openstreetmap.org/ (accessed on 12 April 2020).

applied sciences

MDPI

Article

An Experimental Tuning Approach of Fractional Order Controllers in the Frequency Domain

Isabela Birs [1,*,†], Silviu Folea [1,†], Ovidiu Prodan [2,†], Eva Dulf [1,†] and Cristina Muresan [1,†]

1 Automation Department, Technical University of Cluj-Napoca, Cluj-Napoca 400114, Romania; Silviu.Folea@aut.utcluj.ro (S.F.); Eva.Dulf@aut.utcluj.ro (E.D.); Cristina.Muresan@aut.utcluj.ro (C.M.)
2 Civil Engineering Department, Technical University of Cluj-Napoca, Cluj-Napoca 400114, Romania; Ovidiu.Prodan@mecon.utcluj.ro
* Correspondence: Isabela.Birs@aut.utcluj.ro
† These authors contributed equally to this work.

Received: 10 March 2020; Accepted: 26 March 2020; Published: 31 March 2020

Abstract: Fractional calculus has been used intensely in recent years in control engineering to extend the capabilities of the classical proportional–integral–derivative (PID) controller, but most tuning techniques are based on the model of the process. The paper presents an experimental tuning procedure for fractional-order proportional integral–proportional derivative (PI/PD) and PID-type controllers that eliminates the need of a mathematical model for the process. The tuning procedure consists in recreating the Bode magnitude plot using experimental tests and imposing the desired shape of the closed loop system magnitude. The proposed method is validated in the field of active vibration suppression by using an experimental set-up consisting of a smart beam.

Keywords: fractional calculus; fractional-order control; experimental tuning; smart beam; vibration suppression

1. Introduction

When subjected to the task of improving the performance of a physical process, classical control theory tackles the issue using two steps: mathematically modeling the dynamics of the process and tuning a suitable controller for the identified model.

Several alternatives have been developed for controller tuning based on information gained from the experimental response of the process. For most processes, transient and steady-state analysis of the process offers enough information to obtain a decent controller. However, there is no guarantee that the obtained controller is the ideal one. Experimental methods such as the well-known Ziegler Nichols, Kappa–Tau method, Cohen–Coon , Tyreus–Luyben, Ciancone, and Marlin tune classical proportional–integral–derivative (PID) controllers based on measurements of the different dynamics of the process response [1–5]. In most cases, the controllers tuned using experimental procedures lead to a closed loop system with poor robustness performance [6]. Robust autotuning has been proposed by [7] where the phase and magnitude margins are specified. The problem of autotuning robustness is also tackled in [8] by imposing the iso-damping property of the closed loop.

An increasing trend in controller design integrates the concept of fractional calculus. A rational order for the differentiation and integration extends the classical PID controller to a fractional-order PID. This type of controllers offer increased degrees of freedom and an increased stability as the number of tuned parameters increases [9,10]. One of the popular methods for tuning fractional-order controllers is by imposing frequency domain constraints such as the gain crossover frequency, phase margin, sensitivity, and complementary sensitivity [11]. Other frequency domain tuning procedures may also impose a flat phase around the gain crossover frequency guaranteeing a certain degree of robustness to gain variations [12,13]. Optimization techniques have also been used to successfully tune fractional-order controllers in the frequency domain [14,15]. Over the years, fractional calculus proved useful in both identification and control applications. Its numerous advantages have been successfully tested and validated on a manifold of real-life processes such as flexible transmission [16], active suspension [17], buck converters [18], hydraulic actuators [19], irrigation canals [20], pneumatic pressure [21], industrial distributed systems [22], robotic manipulators [23], etc.

The autotuning approach has also been applied in the fractional calculus controller design [24–27]. A fractional-order experimental tuning procedure based on a single experiment has been explored in [28]. The authors present the tuning of a fractional-order PD controller without using an identified model for the process [29]. The process is fed by a sinusoidal input and the amplitude and phase shifts of the output signal are measured. The output signal is then fed to a filter in order to determine the derivative of the signal. From the obtained information regarding the output amplitude, phase and the derivative slope, a fractional-order PD controller is determined. The Ziegler–Nichols method has been extended in order to tune a fractional-order PI controller in [30]. A fractional-order PID controller is tuned using a relay that takes the system at its stability margin in [31]. For the output signal, the amplitude and phase are measured and using the derivative of the phase the fractional-order PID controller is determined.

The presented work consists of joining the advantages of optimal controller tuning and experimental procedures. An optimal controller tuning in the frequency domain consists of determining the controller parameters such as the shape of the closed loop satisfies several constraints. A fractional-order controller is more suitable for the task, as in the frequency domain, the first-order integration/differentiation introduces a fixed slope of +20/−20 dB/decade and a fixed phase of +90/−90 degrees. A fractional-order integration/differentiation offers the flexibility of variable magnitude slope and a variable phase value. Therefore, a fractional controller is more suitable to honor closed loop frequency domain constraints. The tuning of the fractional-order controller is based on an optimization routine that considers the frequency response magnitude values obtained experimentally.

The main application to the developed method is vibration mitigation. This specific field has gained more popularity in recent years, especially in combination with fractional-order control. The complexity of control tasks associated to vibration suppression systems requires the usage of controllers based on more advanced mathematical tools, such as the fractional calculus branch. Many fractional-order controller tuning methodologies have been developed and tested with the sole purpose of vibration suppression. A fusion between a fractional-order disturbance observer and PI fuzzy controller is proposed in [32] to reduce vibration amplitude in an industrial servo system. Vibrations of an aeroplane wing modeled as a smart beam are suppressed using fractional-order PID-type controllers in [12,13,33–35]. Structural vibrations of uncertain systems are mitigated using a fractional-order sliding mode controller in [36]. Residual vibrations are tackled in [37] using adaptive-predictive fractional-order control strategies, whereas [38] uses optimal fractional-order controllers for AC motor vibrations.

The previously mentioned works use complex fractional-order tuning strategies, based on available models of vibration process' dynamics. The control development is difficult and requires an advanced fractional-order background of the control engineer. In contrast, the experimental fractional-order tuning method presented throughout this study does not require any fractional order background, making the method ideal in industrial applications. From an engineer's point of view, the presented fractional-order controller tuning procedure consists of two steps: The first one implies performing several experiments on the vibration platform, reading the vibration amplitude and experimental approximation of the resonant peak. The second step requires the choice of new magnitude values such as to ensure that the magnitude plot around the magnitude peak is flattened. The ease of application as well as the guaranteed vibration suppression results make the method ideal for industrial scenarios.

The novelty of the presented study is asserted in determining optimal fractional-order controller parameters based solely on the experimental response of the process. The method overcomes drawbacks encountered in classical auto-tuning practices by imposing the magnitude and phase of the closed loop system. A certain degree of robustness can also be assured is the phase margin constraints are imposed to satisfy the iso-damping property. The proposed strategy is appropriate to tune fractional-order proportional derivative (PD), proportional integral (PI), and PID controllers with similar tuning effort.

The developed methodology is validated experimentally on a vibration unit consisting of a smart beam. The movement of the beam is controlled using piezoelectric actuators placed near the fixed end, whereas the displacement is measured with a strain gauge sensor. The choice of the experimental platform lies in its versatility: the smart beam can be correlated to physical systems such as aeroplane wings, helicopter rotors, cantilever beams, suspended platforms, blades of various heavy-duty machinery, etc.

The paper presents the mathematical background of the proposed tuning procedure in Section 2. The method is viable to tune fractional-order PI-, PD-, and PID-type of controllers. Section 3 presents the experimental platform consisting of a smart beam used to prove the efficacy of the proposed method. A fractional-order PD controller is tuned in Section 4, whereas a fractional-order PID is tuned in Section 5. The obtained controllers are validated by analyzing disturbance rejection capabilities.

2. Proposed Tuning Method

The proposed tuning procedure is based on experimental tests without the need of a model for the process. By excluding the model, one can also remove modeling errors and uncertainties, and it also eliminates the need of using a fractional-order model or a high order model in the case of more complex processes. The method has its basis in the Bode magnitude plot which can be created experimentally by feeding the process a series of sine wave signals of a given amplitude A_i and a given frequency ω. As can be seen in Figure 1, for a sinusoidal input, the output of the process will be a sinusoidal output of the same given frequency ω, but with a different amplitude A_o and a phase shift τ.

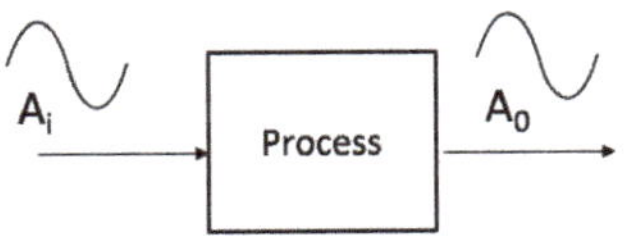

Figure 1. Sinusoidal input and output.

Performing several sine wave tests recreates part of the magnitude frequency diagram through points. The method resides solely on measuring the amplitudes of the input and output signals and computing the corresponding frequency response magnitude values and does not include measurements regarding the time shift, facilitating its ease in real-life applications. The minimum number of points needed resides in the type of the desired controller. Every chosen point of the Bode diagram is part of a system of equations that needs to be solved in order to find the controller parameters, therefore the minimum number of tests needed is equal to the number of controller parameters.

After performing the experimental tests and determining several magnitude values at their corresponding frequencies, the process Bode magnitude plot can be recreated. Next, the shape of the closed loop can be imposed by analyzing the actual values of the open loop magnitude. Taking as an example the vibration phenomenon, which exhibits a resonant characteristic, the magnitude at the resonant frequency should be close to 0, as well as the magnitude at the frequencies near the resonance. In real-life situations, this results in a diminished amplitude of oscillation with less visible effects.

Starting from the following values of the process' magnitude, obtained via the experimental sine tests as described above, for n different frequencies:

$$\begin{cases} |G(\omega_1)| = magG_1\ dB, \\ |G(\omega_2)| = magG_2\ dB, \\ \dots \\ |G(\omega_n)| = magG_n\ dB. \end{cases} \tag{1}$$

The desired magnitude and phase of the closed loop at the chosen frequencies are chosen such as

$$\begin{cases} |H_0(\omega_1)| = magH_1\ dB, & \angle(H_0(\omega_1)) = phaseH1\ deg, \\ |H_0(\omega_2)| = magH_2\ dB, & \angle(H_0(\omega_2)) = phaseH2\ deg, \\ \dots \\ |H_0(\omega_n)| = magH_n\ dB, & \angle(H_0(\omega_n)) = phaseHn\ deg. \end{cases} \tag{2}$$

Choosing the closed loop design specifications at specific frequencies results in shaping the closed loop magnitude and phase diagrams. The main focus of the specifications is to guarantee the disturbance rejection around the resonant frequency by imposing the magnitude value at the resonant frequency close to 0dB. The other magnitude values are chosen such that the slope of the magnitude graph is smoother than the uncompensated system implying an increased damping. The phase constraints are chosen such that the close loop system is stable and a certain degree of robustness is obtained through the iso-damping property.

The closed loop transfer function written depending on the process and the controller for a negative feedback system can be written as

$$H_0(s) = \frac{G(s)C(s)}{1 + G(s)C(s)}, \tag{3}$$

from which the open loop can be deduced as being

$$G(s)C(s) = \frac{H_0(s)}{1 - H_0(s)}. \tag{4}$$

Mapping the Laplace domain to the frequency domain by replacing $s = j\omega$, the frequency domain open loop equation becomes

$$G(j\omega)C(j\omega) = \frac{H_0(j\omega)}{1 - H_0(j\omega)}. \tag{5}$$

Expressing the magnitude of the closed loop in dB is done by

$$|G(j\omega)C(j\omega)|^{dB} = 20\log_{10}\frac{|H_0(j\omega)|}{|1 - H_{0.}(j\omega)|}. \tag{6}$$

The magnitude of the controller in dB is written as

$$|C(j\omega)|^{dB} = |G(j\omega)C(j\omega)|^{dB} - |G(j\omega)|^{dB}. \tag{7}$$

The magnitude of the open loop, denoted by $|G(j\omega)C(j\omega)|$, can be determined from Equation (5) if the magnitude of $1 - H_0(j\omega)$ is determined. In order to do so, the closed loop is written as a sum of its real and imaginary parts, denoted by $A(\omega)$ and $B(\omega)$, respectively.

$$H_0(j\omega) = A(\omega) + jB(\omega). \tag{8}$$

The real and imaginary parts give the magnitude of the open loop

$$|H_0(j\omega)| = \sqrt{A(\omega)^2 + B(\omega)^2} \tag{9}$$

and the phase

$$\angle H_0(j\omega) = \arctan \frac{B(\omega)}{A(\omega)}. \tag{10}$$

Equation (2) imposes values for the closed loop magnitude and phase such that some design specifications are met. Equations (9) and (10) form a system of two equations with two unknowns, $A(\omega)$ and $B(\omega)$. By solving the system created, the values of the real and imaginary parts are obtained. Replacing the magnitudes of $|1 - H_0(j\omega)|$ with the values of $A(\omega)$ and $B(\omega)$ allows the computation of the values for $|G(j\omega)C(j\omega)|^{dB}$ based on Equation (6) and further allows for the estimation of the values of the magnitude of the controller from Equation (7).

Repeating the steps for n test frequencies leads to a system of n equations where the known variable is the magnitude of the controller and the unknowns are the controller parameters.

$$\begin{cases} |C(\omega_1)| = magC_1\, dB, \\ |C(\omega_2)| = magC_2\, dB, \\ \quad \dots \\ |C(\omega_n)| = magC_n\, dB. \end{cases} \tag{11}$$

Customizing the system of equations for the desired type of controller and solving the system should lead to the controller parameters. The system should be solved using numerical optimization routines, such that the best combination of parameters is found. It is highly recommended to take into consideration the following.

- The minimum number of equations needed, denoted by n, depends on the complexity of the chosen controller. The minimum n should be greater or equal than the number of parameters that need to be tuned.
- A constrained routine can be chosen to perform the optimization such that one of the equations from the system is minimized, while the other equations are regarded as constraints.
- As in any optimization algorithm, the developed solution depends on the chosen starting point and the initial points should be chosen realistically. For example, in the case of a fractional-order differentiation, the derivative order belongs to the interval $(0, 1]$.

The proposed tuning method is suitable for tuning classical PID type controllers as well as their fractional-order extensions. A generalization of the PI/PD controller is characterized by the following equation,

$$C_{FO-PI/PD} = k_p + k_d s^{\mu}, \tag{12}$$

where k_p and k_i are the proportional and derivative gains, whereas μ gives the type of the controller,

- $\mu = -1$ integer order PI controller
- $\mu = 1$ integer order PD controller
- $\mu \in (-1, 0)$ fractional-order PI controller
- $\mu \in (0, 1)$ fractional-order PD controller

Choosing this type of controller for the process requires a minimum of three experimental frequency domain tests, as there are three parameters that need to be tuned. If the results obtained by the fractional-order PI/PD controller are not satisfactory, a more complex fractional-order PID may be required:

$$C_{FO-PID} = k_p + k_d s^{\mu} + \frac{k_i}{s^{\lambda}}, \tag{13}$$

where k_p, k_d, and k_i are the proportional, derivative, and integral gains, whereas $\mu \in (0,1]$ denotes the order of the differentiation and $\lambda \in [-1,0)$ expresses the order of the integration. The fractional PID requires a minimum number $n = 5$ of experimental tests in order to develop the system needed for the controller optimization.

The focus of the method is mitigating vibration. The optimization approach is related to the frequency domain magnitude and phase. The method guarantees the computation of a viable fractional-order controller that efficiently suppresses vibration. It is important to emphasize that the efficiency of the obtained fractional-order controller lies in a stable, robust, closed-loop system that reduces the effect of the vibration on the process. The tuning methodology is focused on flattening the resonant peak that causes real-life vibration mitigation, without specific attention to the closed-loop system's performance such as settling time.

As the method is an experimental one and does not require a mathematical model for the process, validating the method will be done on an experimental platform.

3. Experimental Platform

The experimental setup consists of a 250 mm long, 20 mm wide, and 1 mm thick aluminum beam whose vibration is being measured and controlled. The beam is equipped with 4 piezoelectric DuraAct P-878 Power Patch PZT patches, placed two on each side, near the fixed end. Figure 2 presents a 3D model of the set-up. The purpose of two patches from one side is to generate beam oscillations with a given frequency and amplitude, thus giving the possibility to generate a certain disturbance, whereas the other two are used for testing the developed controllers. The fractional controllers are evaluated based on disturbance rejection; the purpose of the controllers being to maintain the free end of the beam from vibrating.

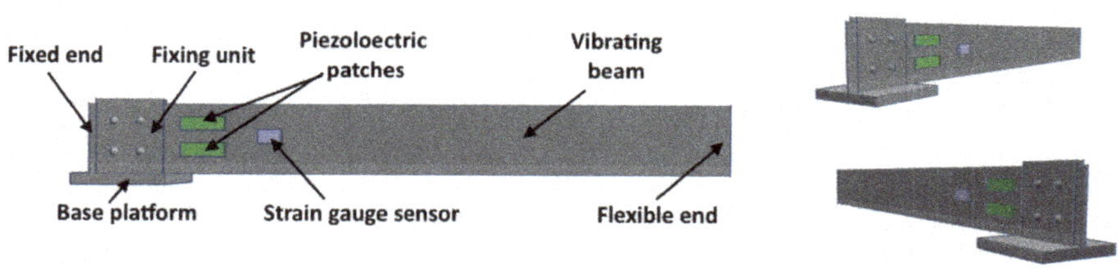

Figure 2. 3D model of the experimental set-up.

The control signal is computed in real-time using $LabVIEW^{TM}$ and is sent to the beam using the CompactRIO^{TM} 9014 controller. The NI 9263 output module excites the PZT patches on the beam, while the NI 9230 input module measures the displacement of the free end using the 120 ohm Omega Prewired KFG-5-120-C1-11L1M2R strain gauge sensors. An additional E-509.X3 module from Physik Instrumente amplifies the signal received from the strain gause sensors, while the E-503.00 amplifies the signal from the PZT patches. A block diagram of the set-up is presented in Figure 3 .

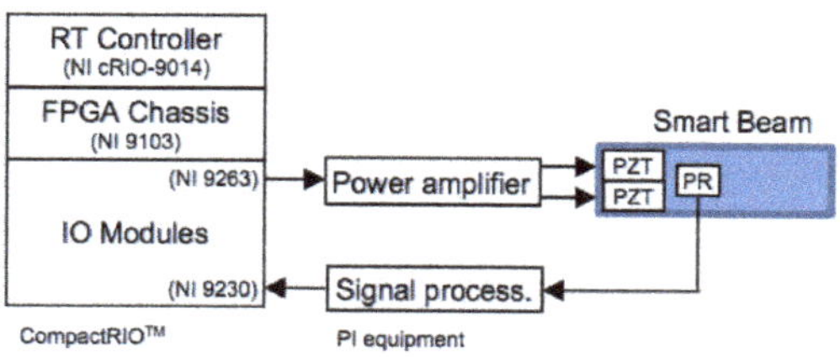

Figure 3. Block diagram of the experimental set-up.

The position of the patches on the beam and a more detailed view of the real workbench is provided in Figure 4.

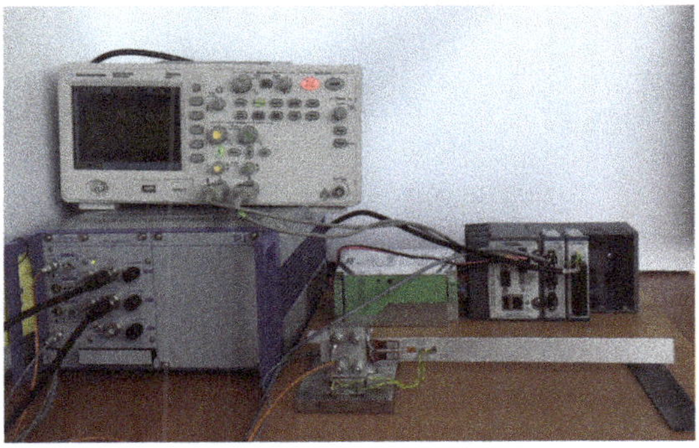

Figure 4. Snapshot of the experimental set-up.

Several experimental tests were performed for different frequencies, and the response of the beam is presented in the figures below. The frequencies chosen are around the resonant frequency, as for this frequency the movement of the beam has the greatest amplitude. The first resonant frequency of the smart beam has been experimentally determined at 14 Hz, as can be seen in Figure 5c.

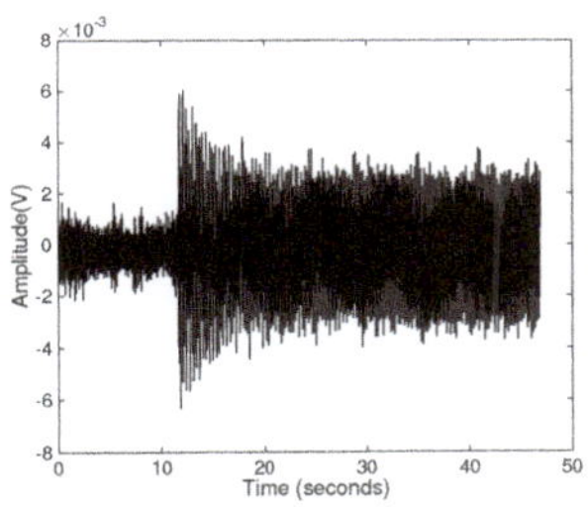

(**a**) Response of the beam to a sine wave of frequency 9 Hz and amplitude 1 V

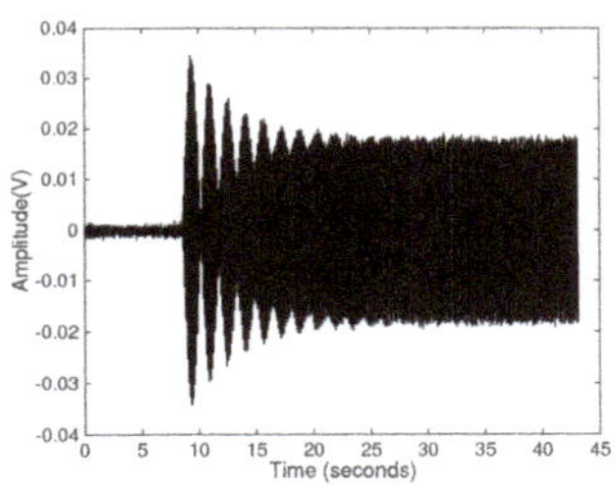

(**b**) Response of the beam to a sine wave of frequency 13 Hz and amplitude 1 V

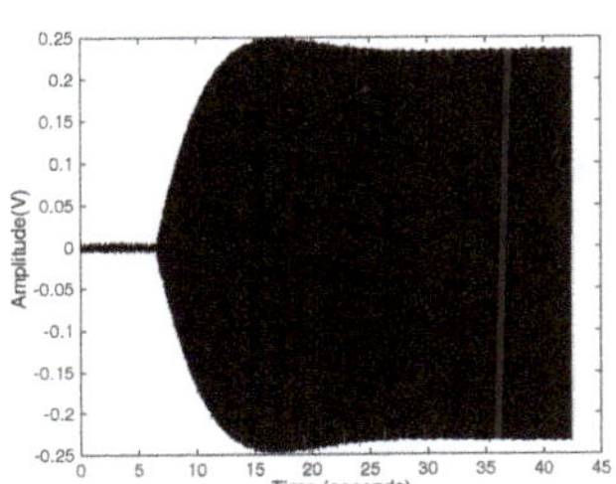

(**c**) Response of the beam to a sine wave of frequency 14 Hz and amplitude 1V

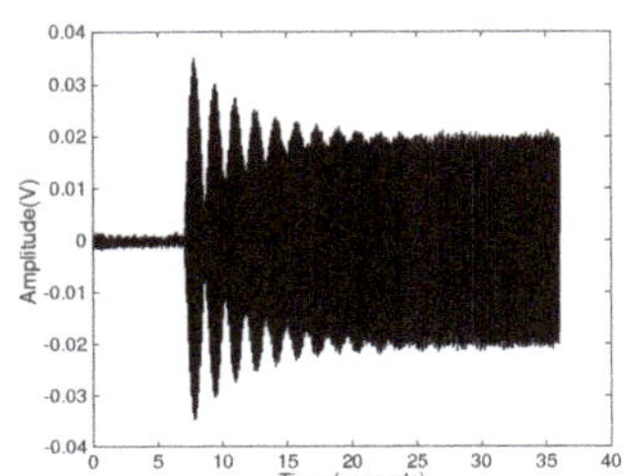

(**d**) Response of the beam to a sine wave of frequency 15 Hz and amplitude 1 V

Figure 5. *Cont.*

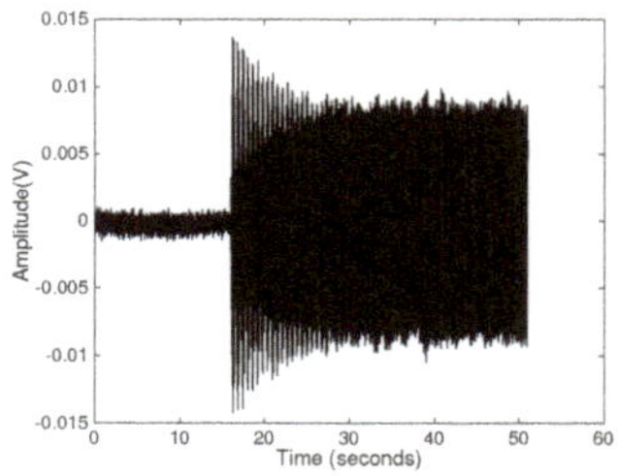

(**e**) Response of the beam to a sine wave of frequency 17 Hz and amplitude 1 V

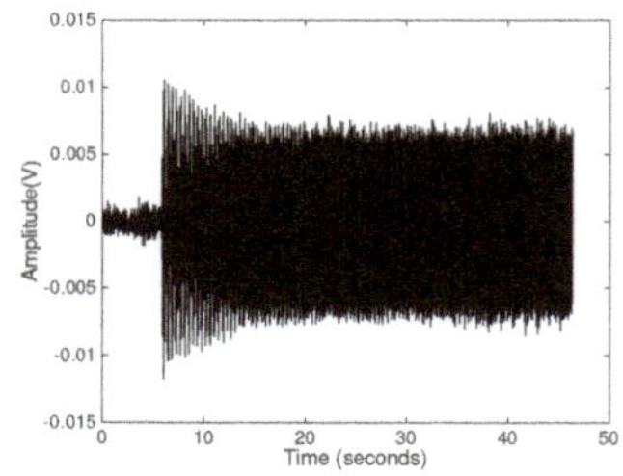

(**f**) Response of the beam to a sine wave of frequency 18 Hz and amplitude 1 V

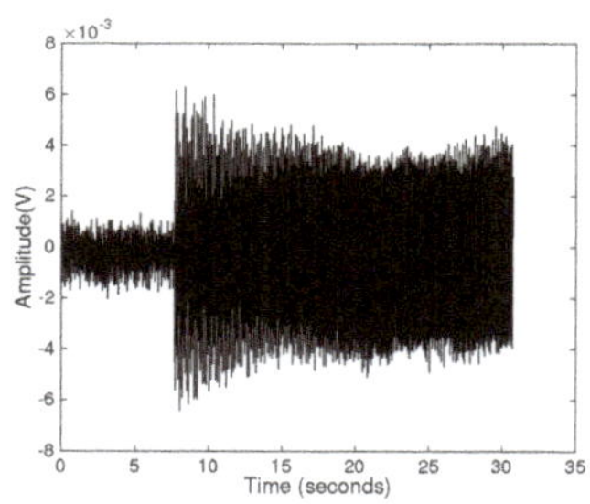

(**g**) Response of the beam to a sine wave of frequency 25 Hz and amplitude 1 V

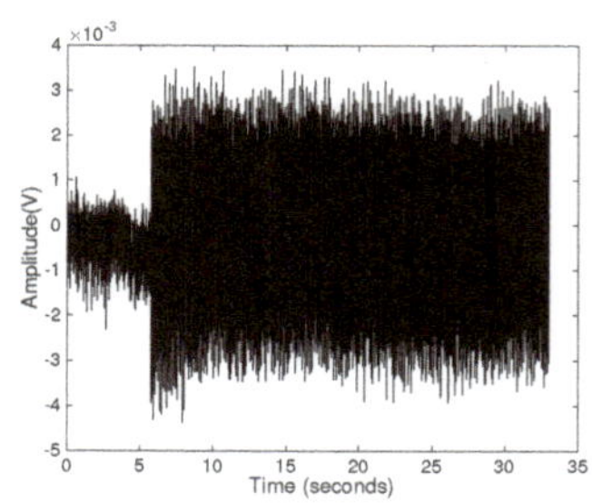

(**h**) Response of the beam to a sine wave of frequency 40 Hz and amplitude 1 V

Figure 5. Experimental response of the beam to sinusoidal inputs of different frequencies.

The Bode magnitude plot was created from data given in Figure 5. Taking as an example Figure 5c, where a sine wave of frequency 14 Hz is given as an input, the measured output amplitude is 0.2355 V. Transforming 14 Hz to rad/s with the formula $f_{rad/s} = f_{Hz} * 2 * \pi$ gives 87.9645 rad/s, whereas 0.2355V to dB using $A_{dB} = 20 * \log_{10} A$ gives -12.5602 dB, giving one point of the Bode diagram from Figure 6. Due to the physical limitations of the test, it is impossible to measure the amplitude for frequencies outside the measured domain because of the low ratio between the measured amplitude and noise sensitivity.

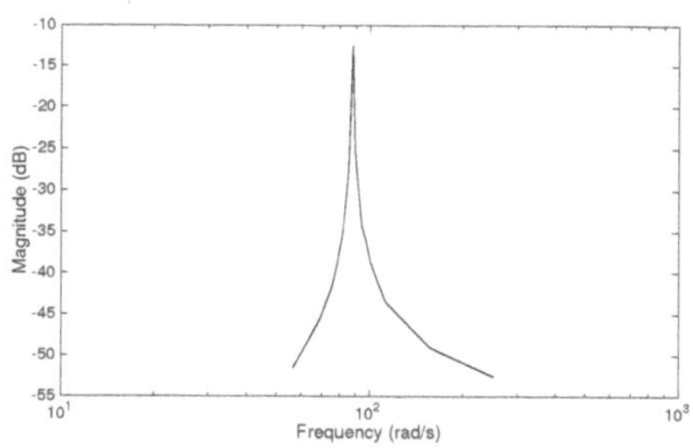

Figure 6. Experimental Bode magnitude plot.

4. Controller Tuning

4.1. Optimization Guidelines

The tuning methodology presented in Section 2 is mainly focused on shaping the Bode magnitude plot. The controller tuning procedure is transformed into a system with multiple nonlinear equations composed of the magnitude of the system at various frequencies. The objective is to find an optimal shape of the magnitude plot that flattens the resonant peak of the system. The optimization can be realized in programs such as MATLAB using functions such as the *fmincon* function provided by the Optimization Toolbox. As in any optimization routine, the starting point for the parameters is paramount in obtaining a valid solution. The proper choice of the pivot point ensures that the provided solution avoids the local minima problem.

When dealing with a fractional-order PID controller, there are five parameters that need to be computed through the optimization procedure: A relevant starting point should be chosen such that the obtained controller is viable in a real life implementation. The fractional orders of differentiation and integration span in the $[0\ \ 1]$ interval. Therefore, any value inside the interval represents a correct choice for the initial condition in the optimization.

However, for the case of the proportional, integral, and derivative gains, the choice for the initial values is more difficult. Illustrating the basic principles on choosing relevant starting points, a fractional-order PD transfer function is considered:

$$H_c(s) = k_p\left(1 + T_d s^{\mu}\right), \tag{14}$$

where k_p is the proportional gain, T_d is the derivative time constant, and μ is the fractional order of differentiation $\mu \in [0\ \ 1]$.

The effect of the fractional-order controller on the magnitude plot will be analyzed further. The magnitude of $H_c(s)$ can be written as $|H_c(j\omega)| = |k_p||1 + T_d(j\omega)^{\mu}|$. The effects of the magnitude of $H_c(j\omega)$ will be analyzed further considering the individual contribution of $|k_p|$ and $|1 + T_d(j\omega)^{\mu}|$.

In an open-loop system, $H_{ol}(j\omega) = H_p(j\omega)H_c(j\omega)$, the magnitude representation of $H_{ol}(j\omega)$ is realized through graphical addition between the magnitude of $H_p(j\omega)$ and $H_c(j\omega)$. The same applies for the individual terms of $H_c(j\omega)$. The proportional gain and derivative time constant are computed based on the assumption that the contribution of every term of the controller is maximum. Looking at the process magnitude plot in Figure 6, the resonant magnitude should be moved closed to 0 dB, according to the methodology presented in Section 2. Therefore, the controller should move the process' magnitude upwards by approximately 15 dB. Assuming that the proportional gain causes the 15 dB movement, one may write that $|k_p| = 15dB = 5.62$. Therefore, any value close to 5.62 may be chosen as the optimization starting point of the proportional gain.

Furthermore, the starting point for the derivative time constant can be approximated using the assumption that $|1 + T_d(j\omega)^{\mu}|$ introduces the 15dB magnitude shift. Using de Moivre's formula, one may write $1 + T_d(j\omega)^{\mu} = 1 + T_d\omega^{\mu}\left(cos\frac{\mu\pi}{2} + jsin\frac{\mu\pi}{2}\right)$. The magnitude of the derivative part can be written as

$$|1 + T_d(j\omega)^{\mu}| = \sqrt{\left(1 + T_d\omega^{\mu}\cos\frac{\mu\pi}{2}\right)^2 + \left(T_d\omega^{\mu} sin\frac{\mu\pi}{2}\right)^2}. \tag{15}$$

Opening the brackets and using the trigonometric property $\sin^2\frac{\mu\pi}{2} + \cos^2\frac{\mu\pi}{2} = 1$ gives

$$|1 + T_d(j\omega)^{\mu}| = \sqrt{1 + 2T_d\omega^{\mu}\cos\frac{\mu\pi}{2} + T_d^2\omega^{2\mu}} = 15dB. \tag{16}$$

Replacing $\omega = \omega_r = 14Hz = 87.9rad/s$, based on Figure 6, leads to

$$|1 + T_d(j\omega_r)^{\mu}| = \sqrt{1 + 2T_d 87.9^{\mu}\cos\frac{\mu\pi}{2} + T_d^2 87.9^{2\mu}} = 5.62. \tag{17}$$

Knowing that $\mu \in [0\ 1]$ and evaluating the solution of (17) for $\mu = 0$ gives $T_{d1} = -175.37$ and $T_{d2} = 0.1744$, whereas for $\mu = 1$, $T_d = \pm 0.0629$. Note that T_d should always be greater than 0. Therefore, the starting point of T_d can be chosen any value around $[0.0629\ \ 0.1744]$.

The guidelines can be extended for the fractional-order PI or PID controller. Note that in the following subsections the PD controller transfer function is slightly different and the derivative gain $k_d = T_d k_p$ interval is $k_d \in [0.3535\ 0.9801]$.

4.2. Fractional Order PI/PD Tuning

A fractional-order PD controller is first designed to suppress unwanted vibrations in the smart beam. As the controller's transfer function has three parameters that have to be computed: k_p, k_d, and μ, three points are computed on the magnitude diagram as follows,

$$\begin{cases} |G(\omega_1)| = -51.52\,dB, & \omega_1 = 56.55\ rad/s, \\ |G(\omega_2)| = -41.25\,dB, & \omega_2 = 107\ rad/s, \\ |G(\omega_3)| = -52.55\,dB, & \omega_3 = 251.3\ rad/s. \end{cases} \tag{18}$$

The three frequencies are chosen as follows; ω_2 is close to the resonant frequency of the beam; and ω_1 and ω_3 are chosen to enclose the resonant frequency, and for the beam case they are chosen as the smallest/greatest frequency for which the magnitude of the process can be computed due to the physical limitations of the experimental set-up. The imposed magnitude and phase closed loop values for the chosen points are

$$\begin{cases} |H_0(\omega_1)| = -12\ dB, & \angle(H_0(\omega_1)) = -0.1\ deg, \\ |H_0(\omega_2)| = -1\ dB, & \angle(H_0(\omega_2)) = -60\ deg, \\ |H_0(\omega_3)| = -20\ dB, & \angle(H_0(\omega_3)) = -120\ deg. \end{cases} \tag{19}$$

The constraints are imposed to ensure a lower resonant peak, larger damping, and a stable closed loop system. The graphical representation of the chosen constraints can be seen in Figure 7.

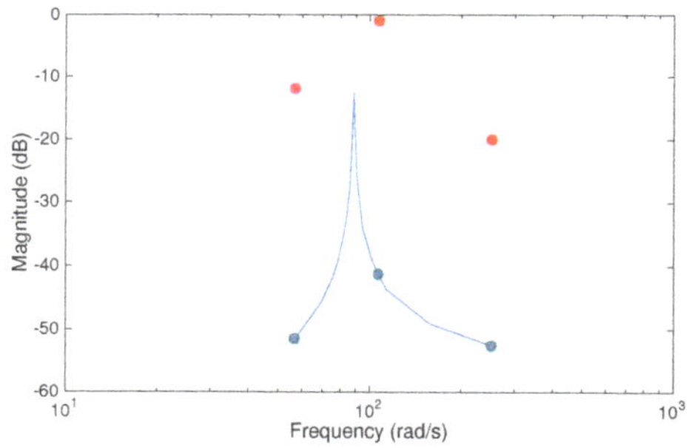

Figure 7. Frequency domain magnitude constraints for fractional-order proportional integral–proportional derivative (PI/PD) tuning (blue—computed magnitude of the open loop system; red—imposed magnitude for the closed loop system).

The optimization routine searches the solution of the minimization problem for $k_p \in (0\ 100)$, $k_d \in (0\ 100)$, and $\mu \in [-1\ 1]$. The initial points chosen for the parameters are $k_p = 15$, $k_d = 0.5$, and $\mu = 0.85$. The controller determined that honors the imposed constraints for the closed loop magnitude and phase is a fractional-order PD controller:

$$C_{FO-PI/PD} = 8 + 0.8888\ s^{0.7683}. \tag{20}$$

The obtained fractional-order PD controller is approximated to a 6th-order discrete transfer function using the method in [39], which is based on mapping the time Laplace operator directly into discrete time. The controller is validated in terms of disturbance rejection.

First, a sinusoidal input of the resonant frequency, 14 Hz, and an amplitude of 1V is given as a disturbance and the response of the beam is presented in Figure 8a. The uncompensated response of the beam leads to an amplitude of 0.25 V. The action of the controller is included at t = 25 s, and the fractional PD controller reduces the oscillation amplitude by 80% with a settling time of 2.5 s. The obtained controller is validated on a larger interval by giving a swept sine input with frequencies between 9 and 90Hz. Two resonant frequencies, representing the first and second flexural modes of the beam, can be visible at 14 Hz and 83 Hz. Figure 8b presents a test performed with the swept sine input for the uncompensated system and two more tests with the control action turned on. It can be seen that the amplitude of the beam is lowered for the entire frequency range.

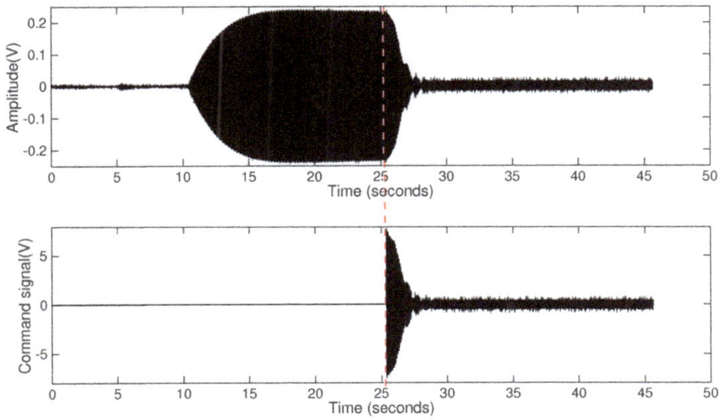

(**a**) Experimental response of the beam when subjected to a sinusoidal disturbance of frequency 14 Hz and amplitude 1 V.

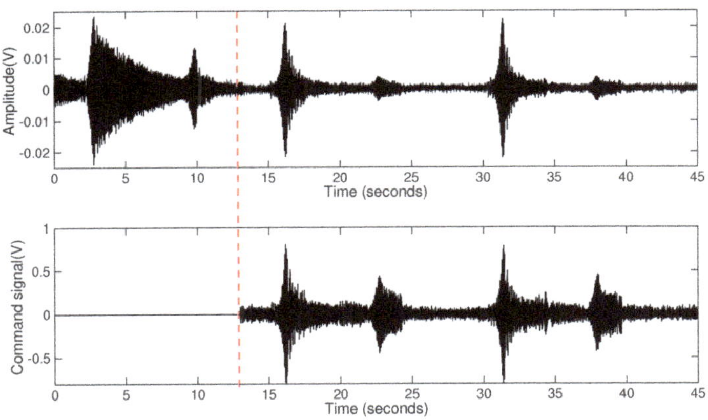

(**b**) Experimental response of the beam when subjected to a swept sine disturbance between 9 Hz and 90 Hz.

Figure 8. Experimental disturbance rejection of the closed loop system with the fractional-order PD controller.

4.3. Fractional-Order PID Tuning

In the previous section, the experimental method was successfully used to tune a fractional-order PD controller. The obtained controller proved useful in the case of disturbance rejection making the controller suitable for the chosen process. However, to prove the efficacy of the proposed method, a more complex fractional-order PID controller is tuned. The transfer function of the fractional PID from (13) has five parameters: k_p, k_i, k_d, λ, and μ. Therefore, a minimum number of five points must be chosen from the magnitude plot.

$$\begin{cases} |G(\omega_1)| = -51.52\, dB, & \omega_1 = 56.55\, rad/s, \\ |G(\omega_2)| = -38.87\, dB, & \omega_2 = 78.54 rad/s, \\ |G(\omega_3)| = -12.56\, dB, & \omega_3 = 87.96\, rad/s, \\ |G(\omega_4)| = -41.25\, dB, & \omega_4 = 107\, rad/s, \\ |G(\omega_5)| = -52.55\, dB, & \omega_5 = 251.3\, rad/s. \end{cases} \tag{21}$$

The frequencies from Equation (17) are chosen as follows; ω_3 is the resonant frequency of the beam as can be seen in Figure 6; ω_1 and ω_5 are lower and upper bounds of the measured frequencies interval, respectively; and ω_2 and ω_4 are evenly distributed frequencies between the resonant peak frequency and the upper and lower bounds, respectively.

The closed loop magnitude and phase are imposed as

$$\begin{cases} |H_0(\omega_1)| = -12\, dB, & \angle(H_0(\omega_1)) = -0.1\, deg, \\ |H_0(\omega_2)| = -2\, dB, & \angle(H_0(\omega_2)) = -60\, deg, \\ |H_0(\omega_3)| = -1\, dB, & \angle(H_0(\omega_3)) = -120\, deg, \\ |H_0(\omega_4)| = -4\, dB, & \angle(H_0(\omega_4)) = -120\, deg, \\ |H_0(\omega_5)| = -20\, dB, & \angle(H_0(\omega_5)) = -120\, deg. \end{cases} \tag{22}$$

Figure 9 presents the chosen values for the magnitude. The magnitude values are chosen in order to reduce the amplitude of the closed loop system at the resonant frequency, therefore ensuring a lower damping. The phase values are chosen such that the closed loop system is stable. The advantage of using a fractional-order PID controller over a fractional-order PD is that robustness can also be a design specification through the iso-damping property. The phase constraints are imposed to be $\angle H_0(\omega_3) = \angle H_0(\omega_4) = \angle H_0(\omega_5) = -120\,\mathrm{deg}$, resulting in a constant phase margin $\phi_m = 60\,\mathrm{deg}$ between the frequencies belonging to the interval $[\omega_3, \omega_5]$, ensuring a certain robustness to small gain variations.

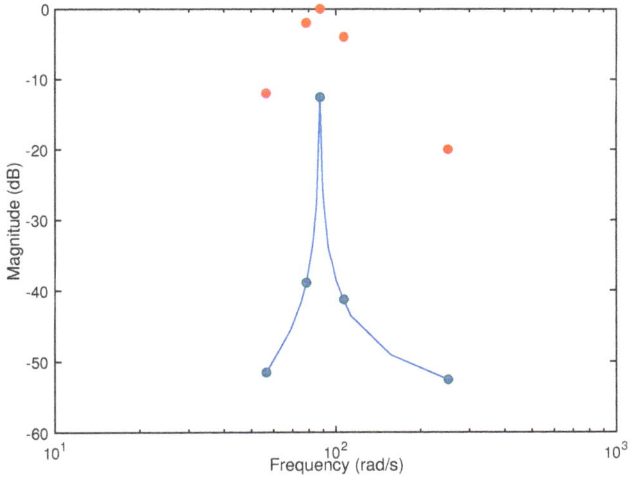

Figure 9. Frequency domain magnitude constraints for fractional-order proportional–integral–derivative (PID) tuning (blue—experimental magnitude values; red—imposed magnitude values for the closed loop system).

Initial points have been given for the optimization routine consisting of $k_p = 15$, $k_d = 0.05$, $k_i = 0.05$, $\mu = 0.85$ and $\lambda = 0.15$. The obtained fractional-order PID controller by performing the minimization for $k_p \in (0, 100)$, $k_d \in (0, 100)$, $k_i \in (0, 100)$, $\mu \in [0, 1]$, and $\lambda \in [0, 1]$ is given by

$$C_{FO-PID} = 15.4506 + 0.8278s^{0.8220} + \frac{0.1}{s^{0.1049}}. \tag{23}$$

The obtained fractional-order PID controller has been approximated using the same complexity, 6th order, and the same settling time, $T_s = 0.005$ s, as the fractional-order PD controller. Implementing the PID controller on the experimental unit successfully diminishes the effect of the disturbance, as can be seen in Figure 10a,b. For a sinusoidal disturbance of 14 Hz and 1V the oscillation amplitude is reduced by 80% and the settling time is around 3 s. The same tests are performed as in the case of the fractional-order PD controller, but the results are different because the two controllers were tuned with different closed loop magnitude and phase constraints. The purpose of tuning the PID controller is not to outperform the fractional PD, but to prove that the method is applicable to different controllers. The fractional-order PD and PID are different controllers tuned using the same method and validated on the same equipment, under the same conditions.

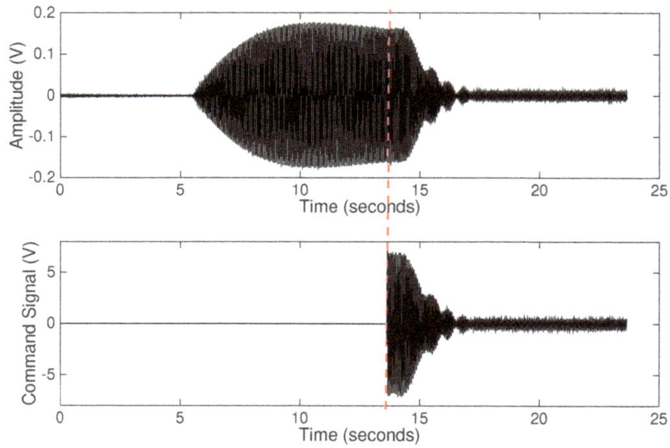

(**a**) Experimental response of the beam when subjected to a sinusoidal disturbance of frequency 14 Hz and amplitude 1 V.

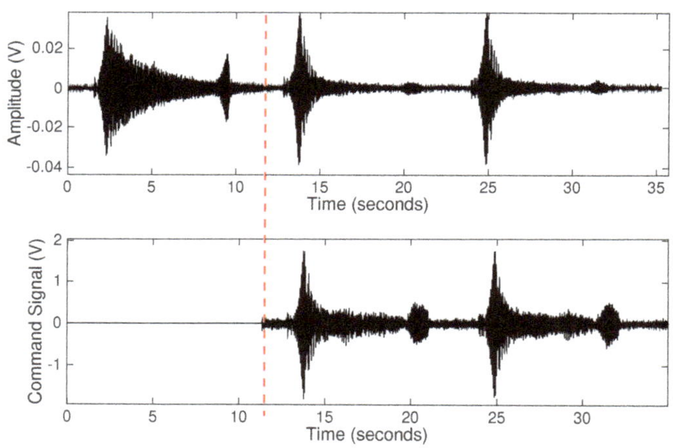

(**b**) Experimental response of the beam when subjected to a swept sine disturbance between 9 Hz and 90 Hz.

Figure 10. Experimental disturbance rejection of the closed loop system with the fractional-order PID controller.

4.4. Robustness Analysis

Robustness to gain variations is indirectly imposed by the constraints for the phase in the interval comprising the resonant frequency. A straight line for the phase in an interval of frequencies ensure that, for small gain variations, the phase margin remains unchanged.

The tuning of the fractional-order PID controller ensures robustness by specifying closed loop phase values in order to honor the iso-damping property. The closed loop phase constraints from Equation (18) strive to achieve a flat phase between $\omega_3 = 87.96\ rad/s$ and $\omega_5 = 251.3\ rad/s$, attaining robustness to small gain variations.

In order to experimentally test the robustness of the obtained controllers from Equations (16) and (19), weights have been added to the free end of the beam such that the resonant frequency is moved from 14 Hz to 12.5 Hz and 11.7 Hz, respectively. Two Neodymium–Iron–Boron permanent magnetic disks with 10mm diameter, 5 mm height, 30 g weight, and 1kg strength were placed on each side of the free end of the beam. The magnets were attached at a 3 cm distance to the moving end, centered with respect to the upper and lower margins of the beam. The tuned fractional-order PD and PID controllers were tested for sinusoidal disturbances of the new resonant frequency.

Figure 11a,b presents the data obtained using the fractional controllers when the resonant frequency of the beam is 12.5 Hz. For the fractional-order PD controller, the settling time is 1.5 s, whereas the PID attenuates and the vibration in approximately 1 s. The oscillation amplitude is reduced by 75% with the fractional-order PD and by 80% with the fractional-order PID controller.

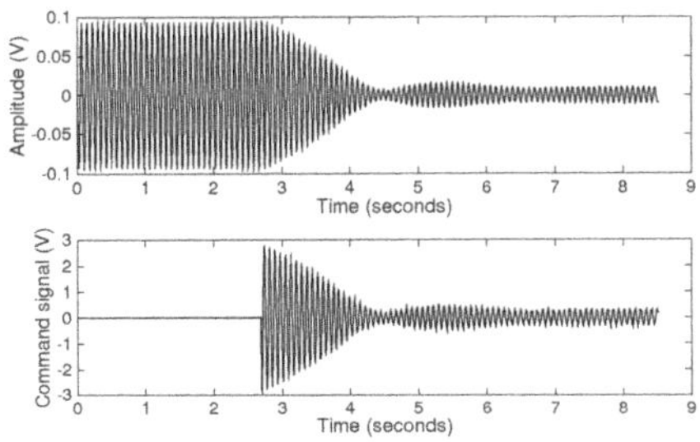

(**a**) Experimental response of the beam with the fractional PD controller when subjected to a sinusoidal disturbance of frequency 12.5 Hz and amplitude 1 V

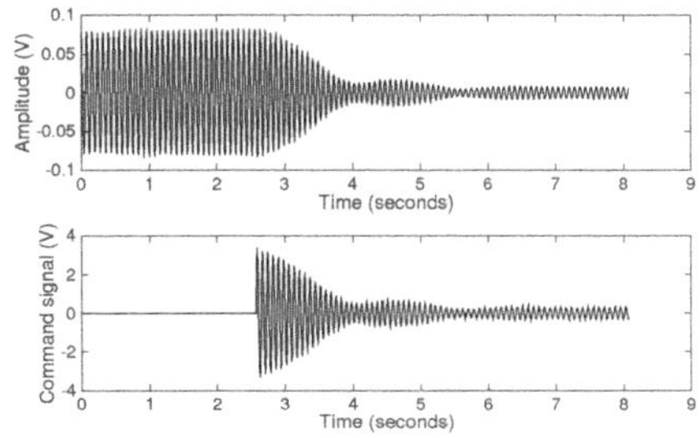

(**b**) Experimental response of the beam with the fractional PID controller when subjected to a sinusoidal disturbance of frequency 12.5 Hz and amplitude 1 V

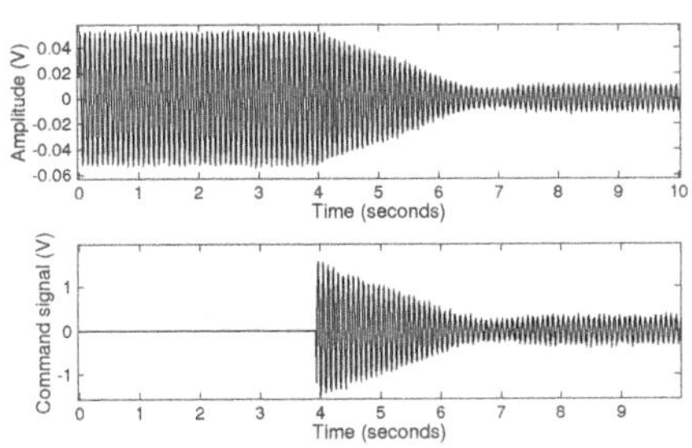

(**c**) Experimental response of the beam with the fractional PD controller when subjected to a sinusoidal disturbance of frequency 11.7 Hz and amplitude 1 V

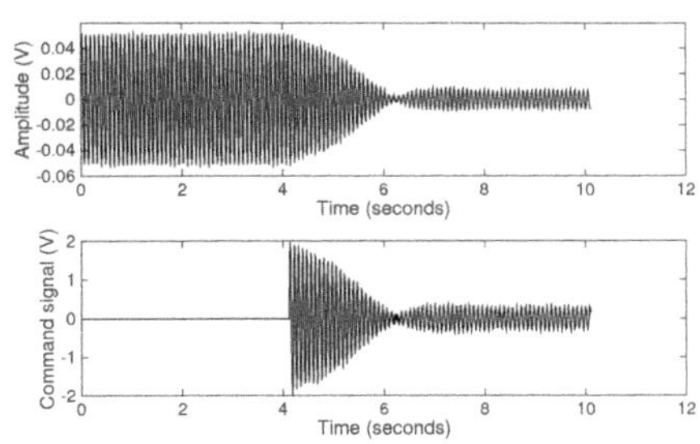

(**d**) Experimental response of the beam with the fractional PID controller when subjected to a sinusoidal disturbance of frequency 11.7 Hz and amplitude 1 V

Figure 11. Experimental robustness validation when altering the resonant frequency of the beam.

Another test is performed when the resonant frequency of the beam is 11.7 Hz and the experimental data obtained with the two controllers is represented in Figure 11c,d. The settling time offered by the fractional-order PD is 2.5 s and the amplitude is reduced by 60%. The fractional-order PID controller reduces the amplitude by 78% in a settling time less than 2 s.

The fractional-order PID overall performance in terms of oscillation amplitude and settling time attenuation is better than the fractional-order PD performance, proving experimentally that the PID controller is more robust. The theoretical explanation is that in the case of the fractional-order PD controller, as there were three constraints imposed, only one constraint was imposed for the phase near the resonant frequency being impossible to impose a robustness condition. However, for the fractional-order PID, three equal phase margin constraints are imposed, leading to a closed loop flat phase around the resonant frequency making the controller robust.

5. Conclusions

The paper presents a tuning procedure for optimal fractional-order controllers based on the frequency domain response. The need of a process model is completely eliminated, making the method easily applicable to complex processes where a higher order model is needed. By eliminating the need of knowing the mathematical representation of the process' dynamics, the tuning procedure eliminates modeling errors.

Most experimental methods are based on measuring both the amplitude and the time shift of the output signal. However, for the chosen process, the presented work has the major advantage of measuring only the output magnitude, completely excluding time shift measurements.

The chosen process for the method validation is a smart beam equipped with piezoelectric actuators. Two fractional-order controllers are tuned with the purpose of diminishing the effect of disturbances on the beam. A fractional-order PD controller and a more complex, fractional-order PID are determined through constrained optimization routines. The tuning process consists first in determining the process frequency response magnitude as some key frequencies based on experimental tests. Next, the closed loop frequency response magnitude and phase values are imposed at these key frequencies, according to some performance specifications. Using these closed loop frequency response magnitude and phase values, the constrained optimization routine is used to estimate the fractional-order controller parameters. The tuning effort is the same for any type of fractional-order controller chosen. The difference between tuning different controllers is the number of chosen points from the frequency diagram as optimization constraints.

The experimental results prove that the proposed method can be successfully used to tune fractional-order controllers in the absence of a process model.

Author Contributions: Conceptualization, I.B., S.F., O.P., E.D., and C.M.; methodology, I.B. and C.M.; software, S.F.; validation, I.B. and S.F.; formal analysis, E.D. and C.M.; investigation, E.. and C.M.; resources, S.F. and O.P.; data curation, I.B. and S.F.; writing—original draft preparation, I.B.; writing—review and editing, C.M. and E.D.; visualization, O.P.; supervision, C.M.; project administration, C.M.; funding acquisition, C.M. All authors have read and agreed to the published version of the manuscript.

Funding: This work was supported by a grant of the Romanian National Authority for Scientific Research and Innovation, CNCS/CCCDI-UEFISCDI, project number PN-III-P1-1.1-TE-2016-1396, TE 65/2018.

Conflicts of Interest: The authors declare no conflicts of interest. The funders had no role in the design of the study; in the collection, analyses, or interpretation of data; in the writing of the manuscript; or in the decision to publish the results.

References

1. Aström, K.J.; Hägglund, T. *PID Controllers: Theory, Design, and Tuning*, 2nd ed.; Instrument Society of America: Research Triangle Park, NC, USA, 1995.
2. de Souza, M.R.S.B.; Murofushi, R.H.; Tavares, J.P.Z.d.S.; Ribeiro, J. Comparison Among Experimental PID Auto Tuning Methods for a Self-balancing Robot. In *Robotics: 12th Latin American Robotics Symposium and Third Brazilian Symposium on Robotics, LARS 2015/SBR 2015, Uberlândia, Brazil, 28 October–1 November 2015*; Revised Selected Papers; Springer International Publishing: Cham, Switzerland, 2016; pp. 72–86. [CrossRef]
3. Keyser, R.D.; Maxim, A.; Copot, C.; Ionescu, C.M. Validation of a multivariable relay-based PID autotuner with specified robustness. In Proceedings of the 2013 IEEE 18th Conference on Emerging Technologies Factory Automation (ETFA), Cagliari, Italy, 10–13 September 2013; pp. 1–6. [CrossRef]
4. Nishikawa, Y.; Sannomiya, N.; Ohta, T.; Tanaka, H. A method for auto-tuning of PID control parameters. *Automatica* **1984**, *20*, 321–332. [CrossRef]
5. Jahanshahi, E.; Sivalingam, S.; Schofield, J.B. Industrial test setup for autotuning of PID controllers in large-scale processes: Applied to Tennessee Eastman process. *IFAC-PapersOnLine* **2015**, *48*, 469–476. [CrossRef]
6. Åström, K.; Hägglund, T. Revisiting the Ziegler–Nichols step response method for PID control. *J. Process Control* **2004**, *14*, 635–650. [CrossRef]

7. Åström, K.J.; Hägglund, T. Automatic tuning of simple regulators with specifications on phase and amplitude margins. *Automatica* **1984**, *20*, 645–651. [CrossRef]
8. Chen, Y.; Moore, K.L. Relay Feedback Tuning of Robust PID Controllers with Iso-damping Property. *Trans. Syst. Man Cyber. Part B* **2005**, *35*, 23–31. [CrossRef] [PubMed]
9. Vilanova, R.; Visioli, A. *PID Control in the Third Millennium: Lessons Learned and New Approaches*; Springer: Berlin, Germany, 2012.
10. Chen, Y.; Petras, I.; Xue, D. Fractional order control—A tutorial. In Proceedings of the 2009 American Control Conference, St. Louis, MO, USA, 10–12 June 2009; pp. 1397–1411. [CrossRef]
11. Monje, C.A. *Fractional-Order Systems and Controls: Fundamentals and Applications*; Springer: London, UK, 2010.
12. Muresan, C.I.; Birs, I.R.; Folea, S.; Dulf, E.H.; Prodan, O. Experimental results of a fractional order PD^{μ} controller for vibration suppresion. In Proceedings of the 2016 14th International Conference on Control, Automation, Robotics and Vision (ICARCV), Phuket, Thailand, 13–15 November 2016; pp. 1–6. [CrossRef]
13. Birs, I.R.; Folea, S.; Copot, D.; Prodan, O.; Muresan, C.I. Comparative analysis and experimental results of advanced control strategies for vibration suppression in aircraft wings. In Proceedings of the 2016 13th European Workshop on advanced Control and Diagnosis, Lille, France, 17–18 November 2016.
14. Birs, I.R.; Muresan, C.I.; Folea, S.; Prodan, O.; Kovacs, L. Vibration suppression with fractional-order $PI^{\lambda}D^{\mu}$ controller. In Proceedings of the 2016 IEEE International Conference on Automation, Quality and Testing, Robotics (AQTR), Cluj-Napoca, Romania, 19–21 May 2016; pp. 1–6. [CrossRef]
15. Birs, I.R.; Folea, S.; Muresan, C.I. An optimal fractional-order controller for vibration attenuation. In Proceedings of the 2017 25th Mediterranean Conference on Control and Automation (MED), Valletta, Malta, 3–6 July 2017; pp. 828–832. [CrossRef]
16. Oustaloup, A.; Mathieu, B.; Lanusse, P. The CRONE Control of Resonant Plants: Application to a Flexible Transmission. *Eur. J. Control* **1995**, *1*, 113–121. [CrossRef]
17. Tseng, H.E.; Hrovat, D. State of the art survey: Active and semi-active suspension control. *Veh. Syst. Dyn.* **2015**, *53*, 1034–1062. [CrossRef]
18. Calderón, A.J.; Vinagre, B.M.; Feliu, V. Fractional order control strategies for power electronic buck converters. *Signal Process.* **2006**, *86*, 2803–2819. [CrossRef]
19. Maddahi, A.; Sepehri, N.; Kinsner, W. Fractional-Order Control of Hydraulically Powered Actuators: Controller Design and Experimental Validation. *IEEE/ASME Trans. Mechatron.* **2019**, *24*, 796–807. [CrossRef]
20. Feliu-Batlle, V.; Rivas-Perez, R.; Castillo-Garcia, F.J.; Sanchez-Rodriguez, L.; Linarez-Saez, A. Robust fractional-order controller for irrigation main canal pools with time-varying dynamical parameters. *Comput. Electron. Agric.* **2011**, *76*, 205–217. [CrossRef]
21. Al-Dhaifallah, M.; Kanagaraj, N.; Nisar, K.S. Fuzzy Fractional-Order PID Controller for Fractional Model of Pneumatic Pressure System. *Math. Probl. Eng.* **2018**, *2018*. [CrossRef]
22. Abdelbaky, M.; El-Hawwary, M.; Emara, H. Implementation of fractional-order PID controller in an industrial distributed control system. In Proceedings of the 2017 14th International Multi-Conference on Systems, Signals & Devices (SSD), Marrakech, Morocco, 28–31 March 2017; pp. 713–718. [CrossRef]
23. Sharma, R.; Rana, K.P.; Kumar, V. Performance analysis of fractional order fuzzy PID controllers applied to a robotic manipulator. *Expert Syst. Appl.* **2014**, *41*, 4274–4289. [CrossRef]
24. Monje, C.A.; Vinagre, B.M.; Feliu, V.; Chen, Y. On auto-tuning of fractional order $PI\lambda D\mu$ controllers. *IFAC Proc. Vol.* **2006**, *39*, 34–39. [CrossRef]
25. Monje, C.; Vinagre, B.; Calderón, A.; Feliu, V.; Chen, Y. Auto-tuning of fractional lead-lag compensators. *IFAC Proc. Vol.* **2005**, *38*, 319–324. [CrossRef]
26. Alagoz, B.B.; Ates, A.; Yeroglu, C. Auto-tuning of PID controller according to fractional-order reference model approximation for DC rotor control. *Mechatronics* **2013**, *23*, 789–797. [CrossRef]
27. Jin, C.Y.; Ryu, K.H.; Sung, S.W.; Lee, J.; Lee, I.B. PID auto-tuning using new model reduction method and explicit PID tuning rule for a fractional order plus time delay model. *J. Process Control* **2014**, *24*, 113–128. [CrossRef]
28. Muresan, C.; Keyser, R.D.; Birs, I.; Folea, S.; Prodan, O. An Autotuning Method for a Fractional Order PD Controller for Vibration Suppression. In Proceedings of the 2017 International Workshop Mathematical Methods in Engineering (MME 2017), Ankara, Turkey, 27–29 April 2017.

29. Keyser, R.D.; Muresan, C.I.; Ionescu, C.M. A novel auto-tuning method for fractional order PI/PD controllers. *ISA Trans.* **2016**, *62*, 268–275. [CrossRef]
30. Gude, J.J.; Kahoraho, E. Modified Ziegler-Nichols method for fractional PI controllers. In Proceedings of the 2010 IEEE 15th Conference on Emerging Technologies Factory Automation (ETFA 2010), Bilbao, Spain, 13–16 September 2010; pp. 1–5. [CrossRef]
31. Monje, C.A.; Vinagre, B.M.; Feliu, V.; Chen, Y. Tuning and auto-tuning of fractional-order controllers for industry applications. *Control Eng. Pract.* **2008**, *16*, 798–812. [CrossRef]
32. Li, W.; Hori, Y. Vibration suppression using single neuron-based PI fuzzy controller and fractional-order disturbance observer. *IEEE Trans. Ind. Electron.* **2007**, *54*, 117–126. [CrossRef]
33. Onat, C.; Şahin, M.; Yaman, Y. Fractional controller design for suppressing smart beam vibrations. *Aircr. Eng. Aerosp. Technol.* **2012**. [CrossRef]
34. Muresan, C.I.; Folea, S.; Birs, I.R.; Ionescu, C. A novel fractional-order model and controller for vibration suppression in flexible smart beam. *Nonlinear Dyn.* **2018**, *93*, 525–541. [CrossRef]
35. Birs, I.R.; Muresan, C.I.; Folea, S.; Prodan, O. A Comparison between Integer and Fractional Order PDμ Controllers for Vibration Suppression. *Appl. Math. Nonlinear Sci.* **2016**, *1*, 273–282. [CrossRef]
36. Aghababa, M.P. A fractional-order controller for vibration suppression of uncertain structures. *ISA Trans.* **2013**, *52*, 881–887. [CrossRef] [PubMed]
37. Lu, J.; Xie, W.; Zhou, H.; Zhang, A. Vibration suppression using fractional-order disturbance observer based adaptive grey predictive controller. *J. Vibroeng.* **2014**, *16*, 2205–2215.
38. Zhao, H.; Deng, W.; Yang, X.; Li, X.; Dong, C. An optimized fractional order PID controller for suppressing vibration of AC motor. *J. Vibroeng.* **2016**, *18*, 2205–2220. [CrossRef]
39. De Keyser, R.; Muresan, C.I.; Ionescu, C.M. An efficient algorithm for low-order direct discrete-time implementation of fractional order transfer functions. *ISA Trans.* **2018**, *74*, 229–238. [CrossRef] [PubMed]

applied sciences

MDPI

Article

Fractional-Order LQR and State Observer for a Fractional-Order Vibratory System

Akihiro Takeshita, Tomohiro Yamashita, Natsuki Kawaguchi and Masaharu Kuroda *

Department of Mechanical Engineering, Graduate School of Engineering, University of Hyogo, 2167 Shosha, Himeji, Hyogo 671-2280, Japan; es19c011@steng.u-hyogo.ac.jp (A.T.); es20o037@steng.u-hyogo.ac.jp (T.Y.); kawaguchi@eng.u-hyogo.ac.jp (N.K.)
* Correspondence: m-kuroda@eng.u-hyogo.ac.jp

Abstract: The present study uses linear quadratic regulator (LQR) theory to control a vibratory system modeled by a fractional-order differential equation. First, as an example of such a vibratory system, a viscoelastically damped structure is selected. Second, a fractional-order LQR is designed for a system in which fractional-order differential terms are contained in the equation of motion. An iteration-based method for solving the algebraic Riccati equation is proposed in order to obtain the feedback gains for the fractional-order LQR. Third, a fractional-order state observer is constructed in order to estimate the states originating from the fractional-order derivative term. Fourth, numerical simulations are presented using a numerical calculation method corresponding to a fractional-order state equation. Finally, the numerical simulation results demonstrate that the fractional-order LQR control can suppress vibrations occurring in the vibratory system with viscoelastic damping.

Keywords: vibration; control; fractional calculus; linear quadratic regulator (LQR); algebraic Riccati equation; iteration; state observer; viscoelasticity

check for updates

Citation: Takeshita, A.; Yamashita, T.; Kawaguchi, N.; Kuroda, M. Fractional-Order LQR and State Observer for a Fractional-Order Vibratory System. *Appl. Sci.* **2021**, *11*, 3252. https://doi.org/10.3390/app11073252

Academic Editors: Cristina I. Muresan and Eva H. Dulf

Received: 18 February 2021
Accepted: 30 March 2021
Published: 5 April 2021

1. Introduction

Fractional calculus is a form of calculus in which the order of differentiation and integration is expanded or generalized from an integer number to a non-integer number. However, fractional calculus is not novel or strange and has a long history that traces back to Leibniz, who was one of the founders of calculus [1]. During these past 30 years, fractional calculus has been applied in engineering fields, especially in control engineering. In control engineering, dynamical systems, which cannot be described sufficiently by conventional integer-order calculus, have been modeled with fractional calculus, and controllers that use fractional calculus have been designed. In other words, a revolution in the construction methods of control system has occurred.

Fractional calculus has been introduced into proportional-integral-differential (PID) control systems as a typical classical control method. Several examples can be cited: $PI^{\lambda}D^{\mu}$ control [2], robust fractional-order PI control using the auto-tuning method [3], the application of this fractional-order PI control method to a manipulator robot [4], fractional-order PD control of a smart beam [5,6], a tuning rule for a fractional-order PD controller for a second-order system [7], fractional-order PID control for a magnetically levitated system [8], fractional-order PD control to suppress vibrations that appear in a three-story structure [9], a graphical design method for tuning a fractional-order PI or PD controller [10], a fractional-order PID controller to suppress vibration that occurs in a cantilever beam using electromagnetic actuators [11], the meaning of the parameters used in fractional-order PID controllers [12], fractional-order PID control for a pumped storage unit [13], fractional-order PID load-frequency control for a power system [14], tuning rules for fractional-order PID controllers [15], $PI^{\lambda}D^{\mu}$ control for trajectory tracking by a 3-DOF parallel manipulator [16], fractional-order PID control for an automatic voltage regulating system [17], and fractional-order PID control for a hydraulic turbine regulating system [18]. Thus, research examples are too numerous to mention.

Fractional calculus has also been used for linear quadratic regulator (LQR) control as a typical modern control theory. In the sense that the controller is applicable to controlled systems with fractional-order dynamics, we use the terminology "fractional-order LQR" in this paper. Similarly, some examples can be cited: a state-feedback LQR controller for fractional-order systems [19], an LQR-based PID controller design using a fractional-order integral performance index [20], a fractional-order LQR [21], a fractional-order LQR control to suppress vibrations appearing in a three-story building [22], and fractional-order PID control for time-delay systems [23]. Again, research examples are too numerous to count. However, it is known that the procedure to derive a fractional-order LQR, as described in a previous report [21], is too complicated to be practical.

Therefore, in the present paper, a novel iteration-based method is first proposed to solve the algebraic Riccati equation to derive the fractional-order LQR. Since all states of the controlled system are fed back in the LQR control scheme, all of the states are assumed to be detected. However, in reality, it is very difficult to measure the states originating from the fractional-order derivative terms. A state observer is then established in order to estimate all of the states, including the fractional-order derivative states. The fractional-order LQR control system is then constructed using the estimated states. Next, by way of example, a viscoelastic damper system modeled with a fractional-order differential equation is considered, and fractional-order LQR control is conducted for the viscoelastic damper system. The effectiveness of the proposed design method is then confirmed. Finally, the proposed fractional-order LQR and the fractional-order LQR derived using the method proposed by Sierociuk and Vinagre [19] are compared with respect to control effects.

The remainder of the present paper is organized as follows. Some basic concepts, such as stability, controllability, and observability for a fractional-order dynamical system, are explained in Section 2. After a viscoelastic damper system is selected as a controlled object in Section 3, the proposed design method for fractional-order LQR control is explained in Section 4. In Section 5, a fractional-order state observer is constructed in order to estimate the fractional-order derivative states, which is required for the realization of fractional-order LQR. A numerical calculation method corresponding to a fractional-order dynamical system is introduced in Section 6. In Section 7, numerical simulation results reveal that the fractional-order LQR, from which feedback gains are derived with the proposed design method, can effectively suppress vibrations that occur in the viscoelastic damper system. Finally, concluding remarks are summarized in Section 8.

2. Background

2.1. Stability of Fractional-Order Dynamical System

Denoting the differential operator as D, a linear fractional-order autonomous system can be given as follows:

$$D^{\alpha}\boldsymbol{x}(t) = \boldsymbol{A}\boldsymbol{x}(t),\ \boldsymbol{x}(0) = \boldsymbol{x}_0, \tag{1}$$

where α is the non-integer order of differentiation, $\boldsymbol{x}(t)$ is the state vector and $\boldsymbol{A}$ is the system matrix.

The dynamical behavior and stability of a fractional-order dynamical system can be analyzed by investigating the poles of the system on the complex plane (s-plane). However, in the case of a fractional-order dynamical system, such as Equation (1), the poles need to be investigated on a new s^{α}-plane. In the present paper, the s^{α}-plane is called the ω-plane. Here, ω can be defined as follows:

$$\omega = s^{\alpha}, \tag{2}$$

where s is given as:

$$s = re^{i\theta}. \tag{3}$$

Then, the definitions given by Equations (2) and (3) yield the following:

$$\omega = s^{\alpha} = \left(re^{i\theta}\right)^{\alpha} = r^{\alpha}e^{i\alpha\theta}. \tag{4}$$

Using Equation (4), lines on the s-plane can be mapped to corresponding lines on the ω-plane. Practically, let us consider the case in which $\alpha = 0.5$. The imaginary axis on the s-plane is expressed as $\theta = \pm\pi/2$, and $0 < r < +\infty$. Then, the imaginary axis on the ω-plane can be expressed as follows:

$$\omega = r^{\frac{1}{2}}e^{\pm i\frac{\pi}{4}}. \tag{5}$$

Furthermore, the negative part of the real axis on the s-plane is expressed as $\theta = \pm\pi$, and $0 < r < +\infty$. Then, the negative part of the real axis on the ω-plane can be expressed as follows:

$$\omega = r^{\frac{1}{2}}e^{\pm i\frac{\pi}{2}}. \tag{6}$$

Thus, the imaginary axis and the negative part of the real axis can be mapped on the ω-plane as illustrated in Figure 1.

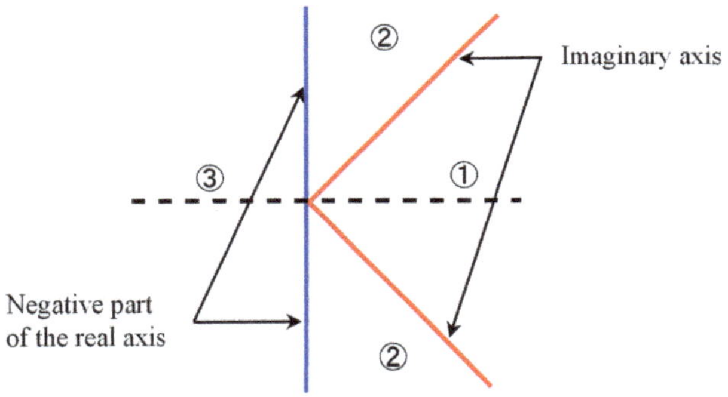

Figure 1. Imaginary axis (red line) and negative part of real axis (blue line) on ω-plane ($\alpha = \frac{1}{2}$).

Classification of the dynamical behavior and stability in each region in Figure 1 is as follows [24]: unstable in region ①, damped oscillation in region ②, and overdamping in region ③. Therefore, the necessary and sufficient condition for the linear autonomous system given in Equation (1) to be asymptotically stable for an arbitrary initial value x_0 is that the arguments of the eigenvalues λ_i of the system matrix A satisfy the following inequality [25,26]:

$$|\arg(\lambda_i)| > \frac{\alpha\pi}{2},\ (i = 1,\ 2,\ \cdots n), \tag{7}$$

where n is the order of the system.

2.2. Controllability of Fractional-Order Dynamical System

Next, a linear fractional-order non-autonomous system can be given as follows:

$$\begin{cases} D^{\alpha}x(t) = Ax(t) + Bu(t), \quad x(0) = x_0 \\ y(t) = Cx(t) \end{cases}, \tag{8}$$

where α designates the non-integer order of differentiation, $x(t)$ is the state vector, $u(t)$ is the input, $y(t)$ is the output, A is the system matrix, B is the input matrix, and C is the output matrix.

For an arbitrary initial state $x(0) = x_0$, a time $t_f > 0$, and a desired value x_f, if an input $u(t)$ that enables a solution to satisfy $x(t_f) = x_f$ exists, then the system described by Equation (8) is controllable. In addition, the propositions shown below are equivalent [25,26].

1. A pair of matrices $(\boldsymbol{A}, \boldsymbol{B})$ is controllable.
2. The rank of the following controllability matrix $\boldsymbol{N}_c$ is n:

$$\boldsymbol{N}_c = \left[\boldsymbol{B}\,\boldsymbol{A}\boldsymbol{B}\,\cdots\,\boldsymbol{A}^{n-1}\boldsymbol{B}\right]. \tag{9}$$

2.3. Observability of Fractional-Order Dynamical System

If the initial state $\boldsymbol{x}(0) = \boldsymbol{x}_0$ can be determined based on the time responses of the input $u(t)$ and the output $y(t)$ at an arbitrary time $t_1 > 0$, then the system defined by Equation (8) is observable. In addition, the propositions given below are equivalent [25,26].

1. A pair of matrices $(\boldsymbol{C}, \boldsymbol{A})$ is observable.
2. The rank of the following observability matrix $\boldsymbol{N}_o$ is n:

$$\boldsymbol{N}_o = \left[\boldsymbol{C}^T\,(\boldsymbol{C}\boldsymbol{A})^T\,\cdots\,\left(\boldsymbol{C}\boldsymbol{A}^{n-1}\right)^T\right]^T. \tag{10}$$

3. Fractional-Order Vibratory System with Viscoelasticity

First, a viscoelastically damped structure is considered as an example of a fractional-order vibratory system in order to explain LQR control. The equation of motion for forced vibration of a one degree-of-freedom viscoelastically damped system with a fractional derivative is:

$$\left[mD^2 + cD^q + kD^0\right]x(t) = f(t), \tag{11}$$

where $x(t)$ is the displacement, m is the mass, c is the viscoelastic damping coefficient, k is the spring constant, t is time, and $f(t)$ is the external force. Dividing both sides of Equation (11) by m leads to the following normal form:

$$\left[D^2 + 2\zeta{\omega_n}^{2-q}D^q + {\omega_n}^2D^0\right]x(t) = u(t), \tag{12}$$

where ${\omega_n}^2 = \frac{k}{m}$, $\zeta = \frac{c}{2m\omega_n^{2-q}}$ and $u(t) = \frac{1}{m}f(t)$.

It is known that q has a value of approximately 0.5 for most viscoelastic materials [27], and is therefore fixed at 0.5 in this paper. Consequently, the viscoelastic damping model can be expressed using a one degree-of-freedom equation of motion known as the generalized Voigt model [28]:

$$\left[D^2 + 2\zeta{\omega_n}^{\frac{3}{2}}D^{\frac{1}{2}} + {\omega_n}^2D^0\right]x(t) = u(t), \tag{13}$$

where $u(t)$ is the control input.

Next, based on the above equation, a fractional-order state equation is derived. Using a sequence of fractional differentiations with a difference in order of 1/2, the state vector $\boldsymbol{x}(t)$ can be defined as follows:

$$\boldsymbol{x}(\mathrm{t}) = \begin{bmatrix} x(t) & D^{\frac{1}{2}}x(t) & D^{\frac{2}{2}}x(t) & D^{\frac{3}{2}}x(t) \end{bmatrix}^T = \begin{bmatrix} x_1(t) & x_2(t) & x_3(t) & x_4(t) \end{bmatrix}^T. \tag{14}$$

Consequently, Equations (13) and (14) lead to the following fractional-order state equation:

$$D^{\frac{1}{2}}\boldsymbol{x}(t) = \boldsymbol{A}\boldsymbol{x}(t) + \boldsymbol{B}u(t), \tag{15}$$

where

$$\boldsymbol{A} = \begin{bmatrix} 0 & 1 & 0 & 0 \\ 0 & 0 & 1 & 0 \\ 0 & 0 & 0 & 1 \\ -\omega_n^2 & -2\zeta\omega_n^{\frac{3}{2}} & 0 & 0 \end{bmatrix}, \boldsymbol{B} = \begin{bmatrix} 0 \\ 0 \\ 0 \\ 1 \end{bmatrix}. \tag{16}$$

4. Fractional-Order LQR Method for State Feedback Control

4.1. Design Method for Fractional-Order LQR

In this section, a fractional-order LQR is designed for the fractional-order vibratory system described in the previous section. The control input is assumed to be given as follows:

$$u(t) = -\boldsymbol{F}\boldsymbol{x}(t), \tag{17}$$

where $\boldsymbol{F}$ designates the feedback-gain matrix.

First, the fractional-order state equation is forced to transform into an integer-order state equation. Setting the closed-loop system matrix as $\boldsymbol{A}_{cl} = \boldsymbol{A} - \boldsymbol{BF}$, Equation (17) is substituted into Equation (15), and the following relation is obtained:

$$D^{\frac{1}{2}}\boldsymbol{x}(t) = \boldsymbol{A}_{cl}\boldsymbol{x}(t). \tag{18}$$

The operator $D^{\frac{1}{2}}$ is applied to both sides of Equation (18) once, and then the following equation is obtained [29]:

$$D\boldsymbol{x}(t) = \boldsymbol{A}_{cl}^2\boldsymbol{x}(t). \tag{19}$$

Second, a quadratic cost function J is considered for the controlled object modeled by Equation (15):

$$J = \int_0^\infty \left[\boldsymbol{x}(t)^T\boldsymbol{Q}\boldsymbol{x}(t) + u(t)^T\boldsymbol{R}u(t)\right]dt, \tag{20}$$

where the weighting matrices $\boldsymbol{Q} \in \mathbb{R}^{n\times n}$ and $\boldsymbol{R} \in \mathbb{R}^{r\times r}$ are positive semi-definite and positive definite, respectively, and the pair of matrices $(\boldsymbol{Q}^{\frac{1}{2}}, \boldsymbol{A})$ is observable. Moreover, Equation (17) is substituted into Equation (20) in order to obtain the following:

$$J = \int_0^\infty \boldsymbol{x}(t)^T\left[\boldsymbol{Q} + \boldsymbol{F}^T\boldsymbol{R}\boldsymbol{F}\right]\boldsymbol{x}(t)dt. \tag{21}$$

Modifying Equation (21) using both Equation (19) and the Lyapunov equation yields the following equations:

$$J = \boldsymbol{x}(0)^T\boldsymbol{P}\boldsymbol{x}(0), \tag{22}$$

$$\boldsymbol{P}\boldsymbol{A}_{cl}^2 + \left(\boldsymbol{A}_{cl}^2\right)^T\boldsymbol{P} = -\boldsymbol{Q} - \boldsymbol{F}^T\boldsymbol{R}\boldsymbol{F}, \tag{23}$$

where $\boldsymbol{P}$ is a positive-definite symmetric solution that satisfies the algebraic Riccati equation. Here, using Lagrange's method for undetermined multipliers, matrices $\boldsymbol{P}$ and $\boldsymbol{F}$, which minimize Equation (22), can be obtained under the constraint condition of Equation (23). If Equation (23) is assumed to be expressed as $\boldsymbol{G}(n \times n) = \{g_{ij}\} = \boldsymbol{0}$, then the number of undetermined multipliers is required to be n^2. Therefore, the undetermined multipliers are arranged in the form of $\boldsymbol{K}(n \times n) = \{k_{ij}\}$. As a result, the Lagrangian function can be expressed as follows [29]:

$$L = J + \sum_{i=1}^{n}\sum_{j=1}^{n} k_{ij}g_{ij} = J + tr\left[\boldsymbol{K}^T\boldsymbol{G}\right] = \boldsymbol{x}(0)^T\boldsymbol{P}\boldsymbol{x}(0) + tr\left[\boldsymbol{K}^T\left\{\boldsymbol{P}\boldsymbol{A}_{cl}^2 + \left(\boldsymbol{A}_{cl}^2\right)^T\boldsymbol{P} + \boldsymbol{Q} + \boldsymbol{F}^T\boldsymbol{R}\boldsymbol{F}\right\}\right], \tag{24}$$

where tr designates the trace of the matrix.

Third, the Lagrangian function L is differentiated with respect to the matrices $\boldsymbol{F}$, $\boldsymbol{P}$, and $\boldsymbol{K}$, respectively, in order to obtain each extreme value. Finally, the equations to obtain the matrices $\boldsymbol{F}$, $\boldsymbol{P}$, and $\boldsymbol{K}$ are derived.

However, it has been clarified that the feedback-gain matrix $\boldsymbol{F}$ cannot be obtained based on the above-mentioned method reported in a previous study [29]. This is because the equation resulting from differentiation directly with respect to $\boldsymbol{F}$ becomes too complicated to be appropriate for obtaining the optimal feedback-gain matrix $\boldsymbol{F}_{opt}$.

4.2. Iteration-Based Method for Obtaining Optimal Feedback Gains

Therefore, in order to overcome this difficulty, the approach must be altered to an iteration-based method in order to obtain the optimal feedback-gain matrix $\boldsymbol{F}_{opt}$. To do so, $\boldsymbol{A}_{cl}^2$ in Equation (24) is expressed as follows:

$$\boldsymbol{A}_{cl}^2 = (\boldsymbol{A}-\boldsymbol{B}\boldsymbol{F})^2 = (\boldsymbol{A}-\boldsymbol{B}\boldsymbol{F}_a)(\boldsymbol{A}-\boldsymbol{B}\boldsymbol{F}_b). \tag{25}$$

Here, $\boldsymbol{F}_a$ is properly determined so that all the real parts of the eigenvalues of $\boldsymbol{A}_{cl}^2$ are negative, and $\boldsymbol{F}_b$ is treated as a variable. By setting $\boldsymbol{A}_{cl\_a} = \boldsymbol{A}-\boldsymbol{B}\boldsymbol{F}_a$, Equation (24) can be rewritten using Equation (25) as follows:

$$L = \boldsymbol{x}(0)^T\boldsymbol{P}\boldsymbol{x}(0) + tr\left[\boldsymbol{K}^T\left\{\boldsymbol{P}\boldsymbol{A}_{cl\_a}(\boldsymbol{A}-\boldsymbol{B}\boldsymbol{F}_b) + [\boldsymbol{A}_{cl\_a}(\boldsymbol{A}-\boldsymbol{B}\boldsymbol{F}_b)]^T\boldsymbol{P} + \boldsymbol{Q} + \boldsymbol{F}_b^T\boldsymbol{R}\boldsymbol{F}_b\right\}\right]. \tag{26}$$

As a result, differentiating the above equation with respect to $\boldsymbol{F}_b$, $\boldsymbol{P}$, and $\boldsymbol{K}$, respectively, produces the following relationships [30]:

$$\frac{\partial L}{\partial \boldsymbol{P}} = \boldsymbol{K}(\boldsymbol{A}-\boldsymbol{B}\boldsymbol{F}_b)^T\boldsymbol{A}_{cl\_a}^T + \boldsymbol{A}_{cl\_a}(\boldsymbol{A}-\boldsymbol{B}\boldsymbol{F}_b)\boldsymbol{K} + \boldsymbol{x}(0)\boldsymbol{x}(0)^T = \boldsymbol{0}, \tag{27}$$

$$\frac{\partial L}{\partial \boldsymbol{F}_b} = -(\boldsymbol{P}\boldsymbol{A}_{cl\_a}\boldsymbol{B})^T\boldsymbol{K} - (\boldsymbol{A}_{cl\_a}\boldsymbol{B})^T\boldsymbol{P}\boldsymbol{K}^T + \boldsymbol{R}^T\boldsymbol{F}_b\boldsymbol{K} + \boldsymbol{R}\boldsymbol{F}_b\boldsymbol{K}^T = \boldsymbol{0}, \tag{28}$$

$$\frac{\partial L}{\partial \boldsymbol{K}} = \boldsymbol{P}\boldsymbol{A}_{cl\_a}(\boldsymbol{A}-\boldsymbol{B}\boldsymbol{F}_b) + (\boldsymbol{A}-\boldsymbol{B}\boldsymbol{F}_b)^T\boldsymbol{A}_{cl\_a}^T\boldsymbol{P} + \boldsymbol{Q} + \boldsymbol{F}_b^T\boldsymbol{R}\boldsymbol{F}_b = \boldsymbol{0}. \tag{29}$$

From Equations (27) and (28), the following equation is obtained because the matrices $\boldsymbol{R}$, $\boldsymbol{P}$, and $\boldsymbol{K}$ are symmetric:

$$\boldsymbol{F}_b = \boldsymbol{R}^{-1}\boldsymbol{B}^T\boldsymbol{A}_{cl\_a}^T\boldsymbol{P}. \tag{30}$$

Next, Equations (29) and (30) generate the following algebraic Riccati equation:

$$\boldsymbol{P}\boldsymbol{A}_{cl\_a}\boldsymbol{A} + (\boldsymbol{A}_{cl\_a}\boldsymbol{A})^T\boldsymbol{P} - \boldsymbol{P}\boldsymbol{A}_{cl\_a}\boldsymbol{B}\boldsymbol{R}^{-1}\boldsymbol{B}^T\boldsymbol{A}_{cl\_a}^T\boldsymbol{P} + \boldsymbol{Q} = \boldsymbol{0}. \tag{31}$$

In the end, an iterative method to obtain the optimal feedback-gain matrix $\boldsymbol{F}_{opt}$ can be explained as follows. Here, $\boldsymbol{P}$ can be obtained from Equation (31), and $\boldsymbol{F}_b$ can be determined by substituting the obtained $\boldsymbol{P}$ into Equation (30). Next, the determined $\boldsymbol{F}_b$ is treated as a new $\boldsymbol{F}_a$, and Equations (25)–(31) are repeatedly calculated until the feedback-gain matrix converges to $\boldsymbol{F}_b = \boldsymbol{F}_a$. Eventually, the converged feedback-gain matrix can be considered to be the optimal feedback-gain matrix $\boldsymbol{F}_{opt}$ to minimize the quadratic cost function J.

5. Fractional-Order LQR Method for Output Feedback Control

5.1. Necessity of Fractional-Order State Observer

The fractional-order LQR control is a kind of state feedback control. Therefore, estimation of the states is an important factor in state feedback control. Numerical simulation of the control may be conducted under the condition that all of the states of the system are assumed to be detected. However, such a condition is not effective in many real cases. Even though sensors are present in the controlled system, obtaining all of the states of the system in most cases is difficult. Furthermore, the viscoelastic damper system, as the controlled object in the present study, contains a fractional-order derivative term that cannot be detected with ordinary sensors in general. Therefore, in order to carry out fractional-order LQR control in practice, it is necessary to estimate the fractional-order derivative states based on information about the observable states.

Accordingly, a scheme that enables estimation of the remaining states based on the input signals and the output signals measured by the sensors is called a state observer. The state observer can be designed as long as the system is observable. If all of the states of the controlled system are estimated using the state observer, then the state feedback control can be conducted using the estimated states. Differences between the estimated states and the actual states are called estimation errors. When the estimation errors become zero, the estimated states match the actual states. Therefore, the state observer must be constructed so that the estimation errors converge to zero.

The estimation errors for the fractional-order system asymptotically go to zero by means of the following design of the state observer:

$$D^{\alpha}\hat{\boldsymbol{x}}(t) = \boldsymbol{A}\hat{\boldsymbol{x}}(t) + \boldsymbol{B}u(t) + \boldsymbol{H}(y(t) - \boldsymbol{C}\hat{\boldsymbol{x}}(t)), \tag{32}$$

where $\hat{\boldsymbol{x}}(t)$, $\boldsymbol{H}$, and $\boldsymbol{C}$ are the estimated state vector, the state observer gain, and the output matrix for the controlled system, respectively. The state observer gain $\boldsymbol{H}$ is selected in order to satisfy the following inequality [31]:

$$|\arg(\lambda_{so\_i})| > \frac{\alpha\pi}{2}, \ (i = 1,\ 2,\ \cdots n), \tag{33}$$

where λ_{so_i} designates an eigenvalue of the matrix $\boldsymbol{A}_{so} = \boldsymbol{A} - \boldsymbol{HC}$.

5.2. Configuration of Fractional-Order State Observer

Next, using Equation (32), a state observer is configured for the viscoelastic damper system. First, the viscoelastic damper, as the controlled object, is assumed to be given by the state equation of Equations (15) and (16). In addition, setting $y(t)$ as the observable, the output equation for the viscoelastic damper system is assumed to be expressed as follows:

$$y(t) = \boldsymbol{Cx}(t), \tag{34}$$

where

$$\boldsymbol{C} = \begin{bmatrix} c_1 & c_2 & c_3 & c_4 \end{bmatrix} \tag{35}$$

in which the values of c_1, c_2, c_3, and c_4 are determined by what is observable. Here, if the pair of matrices $(\boldsymbol{C},\ \boldsymbol{A})$ is observable, then a fractional-order state observer takes the following form:

$$D^{0.5}\hat{\boldsymbol{x}}(t) = \boldsymbol{A}\hat{\boldsymbol{x}}(t) + \boldsymbol{B}u(t) + \boldsymbol{H}(y(t) - \boldsymbol{C}\hat{\boldsymbol{x}}(t)), \tag{36}$$

$$u(t) = -\boldsymbol{F}\hat{\boldsymbol{x}}(t). \tag{37}$$

The subtraction of Equation (15) from Equation (36) yields the fractional-order differential equation for the estimation error $\boldsymbol{e}(t)$:

$$D^{0.5}\boldsymbol{e}(t) = (\boldsymbol{A} - \boldsymbol{HC})\boldsymbol{e}(t), \tag{38}$$

where

$$\boldsymbol{e}(t) = \hat{\boldsymbol{x}}(t) - \boldsymbol{x}(t). \tag{39}$$

From Equation (38), the poles of the matrix $(\boldsymbol{A} - \boldsymbol{HC})$ can be moved freely by the state observer gain $\boldsymbol{H}$, and the convergence characteristics of the estimation error $\boldsymbol{e}(t)$ can be determined freely using the pole assignment method.

Moreover, Equations (15), (34), (36), (37), and (39) produce the following relationships:

$$\begin{aligned} D^{\frac{1}{2}}\boldsymbol{x}(t) &= \boldsymbol{Ax}(t) + \boldsymbol{B}u(t) = \boldsymbol{Ax}(t) + \boldsymbol{B}(-\boldsymbol{F}\hat{\boldsymbol{x}}) = \boldsymbol{Ax}(t) - \boldsymbol{BF}(\boldsymbol{e}(t) + \boldsymbol{x}(t)) \\ &= (\boldsymbol{A} - \boldsymbol{BF})\boldsymbol{x}(t) - \boldsymbol{BFe}(t), \end{aligned} \tag{40}$$

$$\begin{aligned} D^{\frac{1}{2}}\boldsymbol{e}(t) &= D^{\frac{1}{2}}\hat{\boldsymbol{x}}(t) - D^{\frac{1}{2}}\boldsymbol{x}(t) = \boldsymbol{A}\hat{\boldsymbol{x}}(t) + \boldsymbol{B}u(t) + \boldsymbol{H}(y(t) - \boldsymbol{C}\hat{\boldsymbol{x}}) - \boldsymbol{Ax}(t) - \boldsymbol{B}u(t) \\ &= \boldsymbol{A}(\hat{\boldsymbol{x}}(t) - \boldsymbol{x}(t)) - \boldsymbol{H}(\boldsymbol{C}\hat{\boldsymbol{x}}(t) - \boldsymbol{Cx}(t)) = (\boldsymbol{A} - \boldsymbol{HC})(\hat{\boldsymbol{x}}(t) - \boldsymbol{x}(t)) \\ &= (\boldsymbol{A} - \boldsymbol{HC})\boldsymbol{e}(t). \end{aligned} \tag{41}$$

Equations (40) and (41) generate the state equation for the augmented system that includes the state observer as follows [32]:

$$\begin{bmatrix} D^{\frac{1}{2}}\boldsymbol{x}(t) \\ D^{\frac{1}{2}}\boldsymbol{e}(t) \end{bmatrix} = \begin{bmatrix} \boldsymbol{A}-\boldsymbol{BF} & -\boldsymbol{BF} \\ \boldsymbol{0} & \boldsymbol{A}-\boldsymbol{HC} \end{bmatrix} \begin{bmatrix} \boldsymbol{x}(t) \\ \boldsymbol{e}(t) \end{bmatrix}. \tag{42}$$

Figure 2 is a block diagram of the augmented system, in which the state observer is incorporated with the feedback control system.

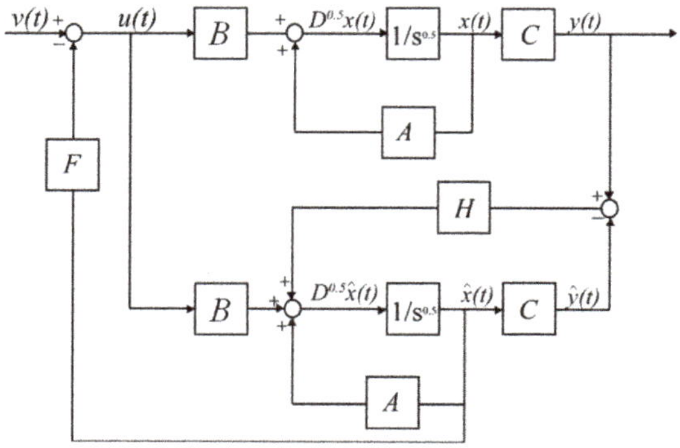

Figure 2. Block diagram of augmented system.

6. Numerical Calculation Method for Fractional-Order Dynamical System

6.1. Numerical Solution of Fractional-Order State Equation

Since a fractional-order state equation contains fractional-order derivative terms, normal numerical calculation methods, such as the Euler method, cannot be used to simulate this equation. Therefore, several numerical calculation approaches corresponding to a differential equation and a state equation that include fractional-order derivative terms have already been proposed [33–36]. One of the numerical calculation methods corresponding to a fractional-order state equation is introduced in this section [35,36].

First, the Grünwald–Letnikov definition, which is a fractional-order differentiation, is as follows:

$${}_aD_t^q f(t) = \lim_{\Delta t \to 0} (\Delta t)^{-q} \sum_{j=0}^{[\frac{t-a}{\Delta t}]} (-1)^j \begin{pmatrix} q \\ j \end{pmatrix} f(t - j\Delta t), \tag{43}$$

where Δt is the pitch width, q is the fractional-order of differentiation, $[\,]$ is the integer part of the argument, and $\begin{pmatrix} q \\ j \end{pmatrix}$ indicates the generalized binomial coefficient. The revised version of the above definition for the numerical calculation is as follows:

$${}_aD_t^q f(t) = \lim_{\Delta t \to 0} \frac{1}{(\Delta t)^q} \sum_{j=0}^{[\frac{t-a}{\Delta t}]} w_j^{(q)} f(t - j\Delta t), \tag{44}$$

where

$$w_0^{(q)} = 1,\ w_j^{(q)} = \left(1 - \frac{q+1}{j}\right) w_{j-1}^{(q)},\ j = 1,\ 2,\ \cdots. \tag{45}$$

Next, a simple scalar differential equation is considered as follows:

$$D^q z(t_k) = f(t_k,\ z(t_k)), \tag{46}$$

where the number of calculations is k, and the time is set to $t_k = k\Delta t$. Applying Equation (44) to the left-hand side of the above equation yields [36]:

$$\frac{1}{(\Delta t)^q}\sum_{j=0}^{m} w_j^{(q)} z\left(t_{k-j}\right) = \frac{1}{(\Delta t)^q}\left[z(t_k) + \sum_{j=1}^{m} w_j^{(q)} z\left(t_{k-j}\right)\right] = f(t_k,\, z(t_k)), \tag{47}$$

where $m = \left[\frac{t_k}{\Delta t}\right] = k$.

Furthermore, Equation (47) is transformed so that this equation can be solved in terms of z, and then the following relation is obtained:

$$z(t_k) = (\Delta t)^q f(t_k,\, z(t_k)) - \sum_{j=1}^{m} w_j^{(q)} z\left(t_{k-j}\right). \tag{48}$$

If the pitch width Δt is sufficiently small, then the first term on the right-hand side of Equation (48) can be approximated as follows [36]:

$$(\Delta t)^q f(t_k,\, z(t_k)) \approx (\Delta t)^q f(t_k,\, z(t_{k-1})). \tag{49}$$

By this approximation, the numerical calculation becomes possible.

Moreover, if each row of the system matrix A of Equation (8) is dealt with as a function $f(t)$ in Equation (46), then the above-mentioned numerical calculation method can be used to solve the fractional-order state equation as follows [36]:

$$\begin{cases} z_1(t_k) = (\Delta t)^q f_1(t_k,\, z_1(t_{k-1}),\, z_2(t_{k-1}), \cdots,\, z_n(t_{k-1})) - \sum\limits_{j=1}^{m} w_j^{(q)} z_1\left(t_{k-j}\right) \\ z_2(t_k) = (\Delta t)^q f_2(t_k,\, z_1(t_k),\, z_2(t_{k-1}), \cdots,\, z_n(t_{k-1})) - \sum\limits_{j=1}^{m} w_j^{(q)} z_2\left(t_{k-j}\right) \\ \vdots \\ z_n(t_k) = (\Delta t)^q f_n(t_k,\, z_1(t_k),\, z_2(t_k), \cdots,\, z_n(t_{k-1})) - \sum\limits_{j=1}^{m} w_j^{(q)} z_n\left(t_{k-j}\right) \end{cases}. \tag{50}$$

In the present paper, each state $x_i(t_k)$ is applied to each $z_i(t_k)$. In the case that the Caputo definition is used, because of the relationship between the Caputo definition and the Grünwald–Letnikov definition, the initial value for each state must be dealt with as follows:

$$z_i(t_k) = x_i(t_k) - x_i(0). \tag{51}$$

Equation (51) is substituted into Equation (50), and then Equation (50) can be rewritten as follows:

$$\begin{cases} x_1(t_k) = (\Delta t)^q f_1(t_k,\, x_1(t_{k-1}),\, x_2(t_{k-1}), \cdots,\, x_n(t_{k-1})) - \sum\limits_{j=1}^{m} w_j^{(q)} \left\{x_1\left(t_{k-j}\right) - x_1(0)\right\} + x_1(0) \\ x_2(t_k) = (\Delta t)^q f_2(t_k,\, x_1(t_k),\, x_2(t_{k-1}), \cdots,\, x_n(t_{k-1})) - \sum\limits_{j=1}^{m} w_j^{(q)} \left\{x_2\left(t_{k-j}\right) - x_2(0)\right\} + x_2(0) \\ \vdots \\ x_n(t_k) = (\Delta t)^q f_n(t_k,\, x_1(t_k),\, x_2(t_k), \cdots,\, x_n(t_{k-1})) - \sum\limits_{j=1}^{m} w_j^{(q)} \left\{x_n\left(t_{k-j}\right) - x_n(0)\right\} + x_n(0) \end{cases}. \tag{52}$$

Moreover, when the amount of past calculation data is large, the calculation of the second term on the right-hand side of each row in Equation (52) becomes large. Therefore, the parameter L_0, which determines the truncated term in the past calculation data, is introduced so that the second term on the right-hand side of each row in Equation (52) is made to be approximated as follows [36]:

$$\sum_{j=1}^{m} w_j^{(q)} \left\{ x_i \left(t_{k-j} \right) - x_i(0) \right\} \approx \sum_{j=1}^{\min(L_0,\, k)} w_j^{(q)} \left\{ x_i \left(t_{k-j} \right) - x_i(0) \right\}. \tag{53}$$

Using the numerical calculation method described above, the fractional-order state equation and the fractional-order state observer can be simulated numerically.

6.2. Comparison between Numerical and Exact Solutions

Next, the solution obtained by the above-mentioned numerical calculation method is compared with the exact solution. The exact solution of Equation (15) can be given as follows using the Mittag–Leffler function $\boldsymbol{E}_{a_1,\, a_2}$ [29,37]:

$$\boldsymbol{x}(t) = \boldsymbol{E}_{\frac{1}{2},\,1}\left(\boldsymbol{A} t^{\frac{1}{2}} \right) \boldsymbol{x}(0) + \int_0^t (t-\tau)^{\frac{1}{2}-1} \boldsymbol{E}_{\frac{1}{2},\,\frac{1}{2}} \left(\boldsymbol{A}(t-\tau)^{\frac{1}{2}} \right) \boldsymbol{B} u(\tau) d\tau, \tag{54}$$

$$\boldsymbol{E}_{a_1,\, a_2}(z) = \sum_{k=0}^{\infty} \frac{z^k}{\Gamma(a_1 k + a_2)},\ (a_1 > 0,\ a_2 > 0). \tag{55}$$

In addition, the parameters for the viscoelastic damper are set to $\zeta = 0.1$ and $\omega_n = 3.0\ \mathrm{rad/s}$. The initial conditions for the states are set to $\boldsymbol{x}(0) = \begin{bmatrix} 0 & 0 & 1 & 0 \end{bmatrix}^T$. For each state, the numerical calculation results and the exact solutions obtained under the free vibration condition are compared in the following figures.

First, Figure 3 shows calculation results for simulation conditions with a larger Δt than those in Figure 4. The numerical calculations in Figure 4 are more accurate than those in Figure 3, because the gap between the numerical calculation results and the exact solution in Figure 4 is smaller than that in Figure 3. Based on this result, it is understood that a smaller Δt improves the accuracy of numerical calculations. A numerical calculation method that enables a highly accurate simulation of a state equation that includes fractional-order derivative terms has thus been clarified.

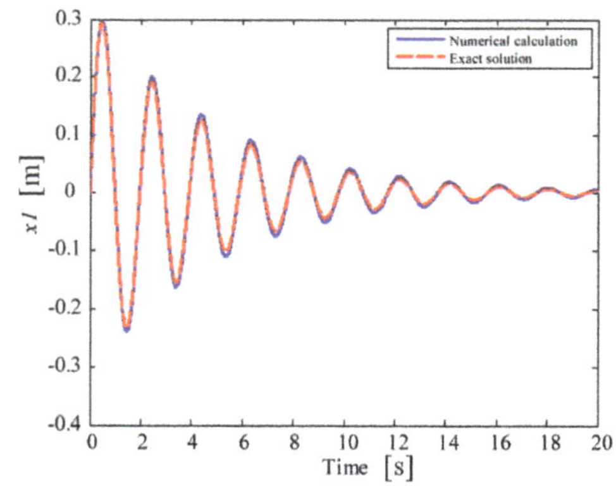

(a) x_1

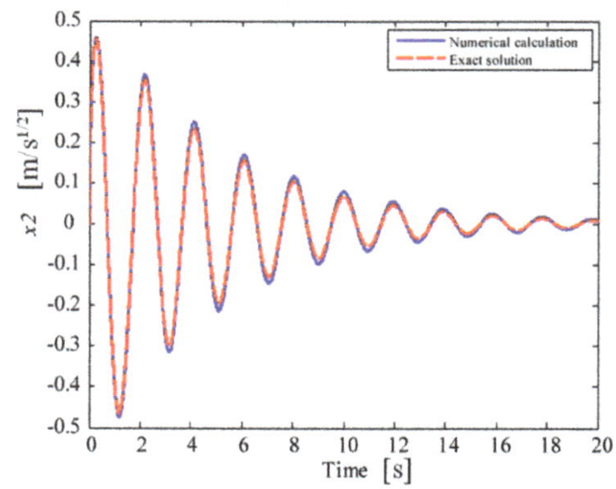

(b) x_2

Figure 3. *Cont.*

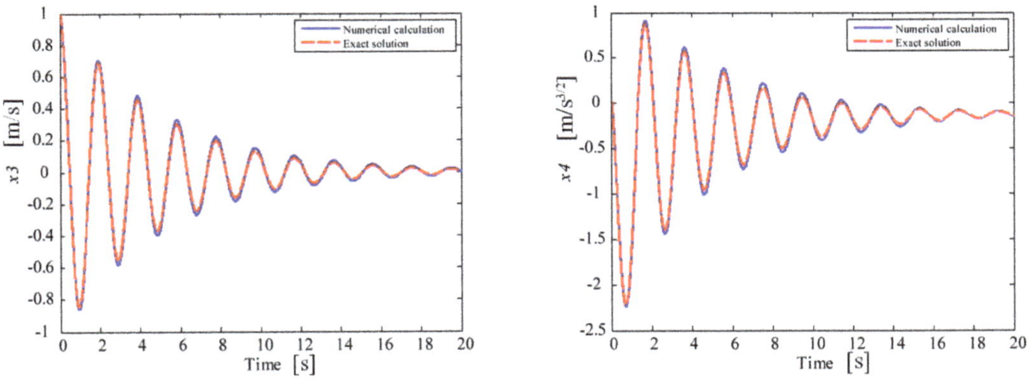

(**c**) x_3 (**d**) x_4

Figure 3. Numerical calculation results (blue lines) and exact solutions (red broken lines) ($\Delta t = 0.001$).

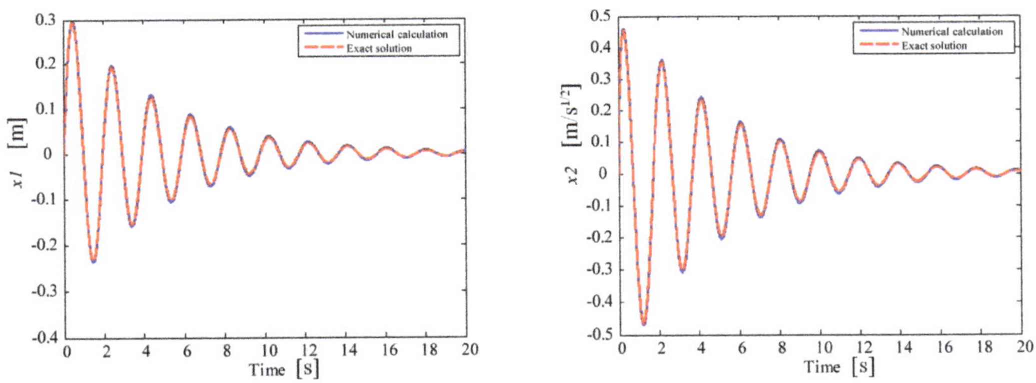

(**a**) x_1 (**b**) x_2

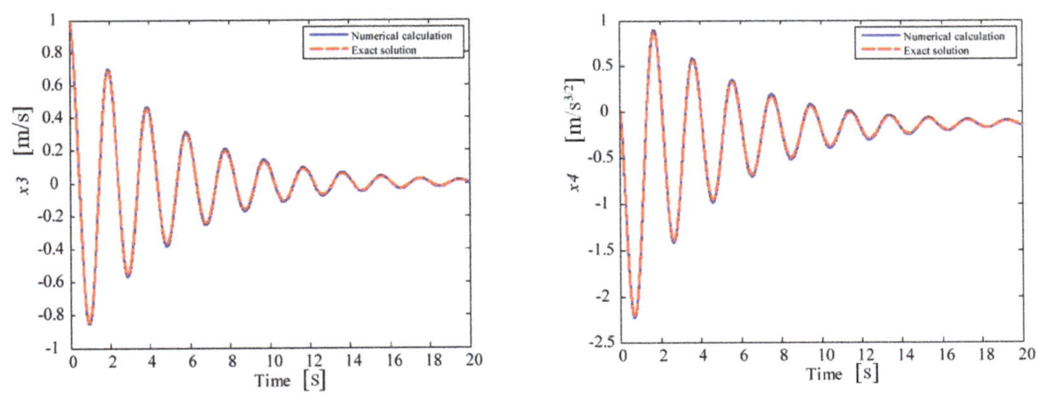

(**c**) x_3 (**d**) x_4

Figure 4. Numerical calculation results (blue lines) and exact solutions (red broken lines) ($\Delta t = 0.0005$).

7. Illustrative Examples

7.1. Fractional-Order LQR Control for Viscoelastic Damper System

In this section, the viscoelastic damper described in Section 3 is set to be the controlled object. Moreover, the feedback gains of the fractional-order LQR are obtained using the method explained in Section 4, and the states are estimated with the fractional-order state observer described in Section 5. In addition, each state is simulated using the numerical calculation method introduced in Section 6. Another derivation method for the feedback gains of the fractional-order LQR, which is different from the method proposed herein, was proposed by Sierociuk and Vinagre [19]. Accordingly, vibration waveforms for the states controlled by the LQR using the feedback gains obtained with the proposed method are compared with those obtained using the Sierociuk and Vinagre method.

The parameters for the viscoelastic damper system are set as $\zeta = 0.1$ and $\omega_n = 3.0$ rad/s. The observable is chosen to be the displacement. Therefore, the output matrix $\boldsymbol{C}$ in Equation (35) is as follows:

$$\boldsymbol{C} = \begin{bmatrix} 1 & 0 & 0 & 0 \end{bmatrix}. \tag{56}$$

First, the stability of the viscoelastic damper system expressed by Equation (15) is investigated. The eigenvalues of the system matrix $\boldsymbol{A}$ are given as:

$$\lambda = 1.2263 \pm 1.3098i,\ -1.2263 \pm 1.1366i. \tag{57}$$

Here, the stability of the system is determined using Equation (7):

$$\frac{2}{\pi}|\arg(\lambda_i)| = 0.5210,\ 0.5210,\ 1.5241,\ 1.5241 > 0.5. \tag{58}$$

As a result of this inequality, this system can be said to be asymptotically stable.

Next, the controllability and the observability of the viscoelastic damper system are examined. The ranks of the matrices $\boldsymbol{N}_c$ and $\boldsymbol{N}_o$ in Equations (9) and (10) can be calculated as follows:

$$rank(\boldsymbol{N}_c) = 4,\ rank(\boldsymbol{N}_o) = 4. \tag{59}$$

The pair of matrices $(\boldsymbol{A}, \boldsymbol{B})$ for the viscoelastic damper system is controllable, and the pair of matrices $(\boldsymbol{C}, \boldsymbol{A})$ for the viscoelastic damper system is observable. Based on these results, the system can be said to be controllable and observable. Therefore, a state feedback controller and a state observer can be designed for this system.

If the weighting matrices $\boldsymbol{Q}$ and $\boldsymbol{R}$ in the quadratic cost function J are designed as follows:

$$\boldsymbol{Q} = \begin{bmatrix} 10 & 0 & 0 & 0 \\ 0 & 0 & 0 & 0 \\ 0 & 0 & 10 & 0 \\ 0 & 0 & 0 & 0 \end{bmatrix},\ \boldsymbol{R} = 1, \tag{60}$$

then the optimal feedback-gain matrix $\boldsymbol{F}_{opt}$ is obtained as follows:

$$\boldsymbol{F}_{opt} = \begin{bmatrix} 0.5394 & -0.5520 & 3.3285 & -0.0098 \end{bmatrix}. \tag{61}$$

In this case, the weights in $\boldsymbol{Q}$ are placed only on the integer-order derivative states. With these feedback gains, the eigenvalues for the closed-loop system-matrix $\boldsymbol{A}_{cl} = \boldsymbol{A} - \boldsymbol{BF}$ are shown below:

$$\lambda_{cl} = 0.8475 \pm 1.5897i,\ -0.8426 \pm 1.4931i. \tag{62}$$

From Equation (62), since the poles of the closed-loop system are located in region ② in Figure 1, the system is asymptotically stable.

Next, a fractional-order state observer is designed using Equation (36) by the pole-assignment method. Furthermore, generally, when a state observer and a regulator are used simultaneously, the state observer must be designed to be able to estimate the states of the system faster than the regulator makes the states converge to zeros. Here, the eigenvalues of the matrix $\boldsymbol{A}_{so} = \boldsymbol{A} - \boldsymbol{HC}$ are designed as follows:

$$\lambda_{so} = -10.0,\ -9.0,\ -8.0,\ -7.0. \tag{63}$$

The stability of these eigenvalues can be determined using Equation (7) as:

$$\frac{2}{\pi}|\arg(\lambda_{cl\_i})| = 0.6882,\ 0.6882,\ 1.3271,\ 1.3271 > 0.5, \tag{64}$$

$$\frac{2}{\pi}|\arg(\lambda_{so\_i})| = 2,\ 2,\ 2,\ 2 > 0.5. \tag{65}$$

The closed-loop system $\boldsymbol{A}_{cl}$ and the state observer $\boldsymbol{A}_{so}$ are asymptotically stable. Judging from Equations (62) and (63), the eigenvalues of the matrix $\boldsymbol{A}_{so}$ are more stable than those of the matrix $\boldsymbol{A}_{cl}$. This proves that state estimation by the state observer can be designed to be faster than state convergence by the regulator. Consequently, the state observer gain matrix $\boldsymbol{H}$ is obtained as follows:

$$\boldsymbol{H} = 10^3 \times \begin{bmatrix} 0.0340 & 0.4310 & 2.4130 & 4.9957 \end{bmatrix}. \tag{66}$$

In practice, in the simulation, the initial conditions for the states and those for the estimated values are assumed to be as follows:

$$\boldsymbol{x}(0) = \begin{bmatrix} 0 & 0 & 1 & 0 \end{bmatrix}^T, \tag{67}$$

$$\hat{\boldsymbol{x}}(0) = \begin{bmatrix} 0 & 0 & 1.2 & 0 \end{bmatrix}^T. \tag{68}$$

In addition, the parameter in the numerical calculation method is set as $\Delta t = 0.0005$.

Using the above-mentioned values, whether the fractional-order state observer can estimate all of the states is first investigated under the free vibration condition. Figure 5 shows each estimation error, and Figure 6 compares each actual state and each estimated state. As shown in Figure 5, each estimation error converges to zero, although the estimation error about x_4 converges more slowly than other errors. Figure 6 demonstrates that each state can be estimated successfully by the fractional-order state observer.

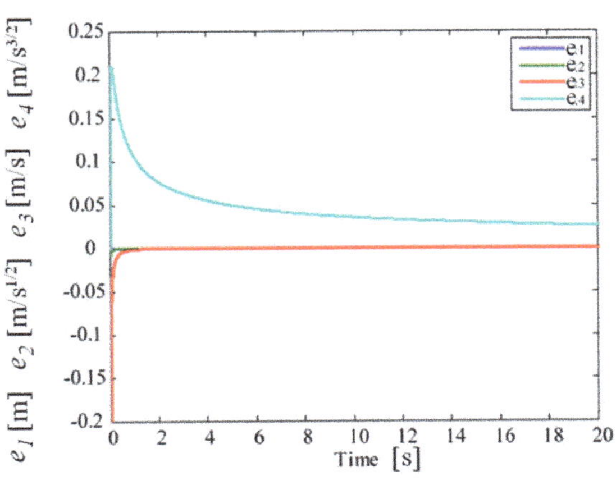

Figure 5. Estimation errors for states (e_1: blue line, e_2: green line, e_3: red line, and e_4: light blue line).

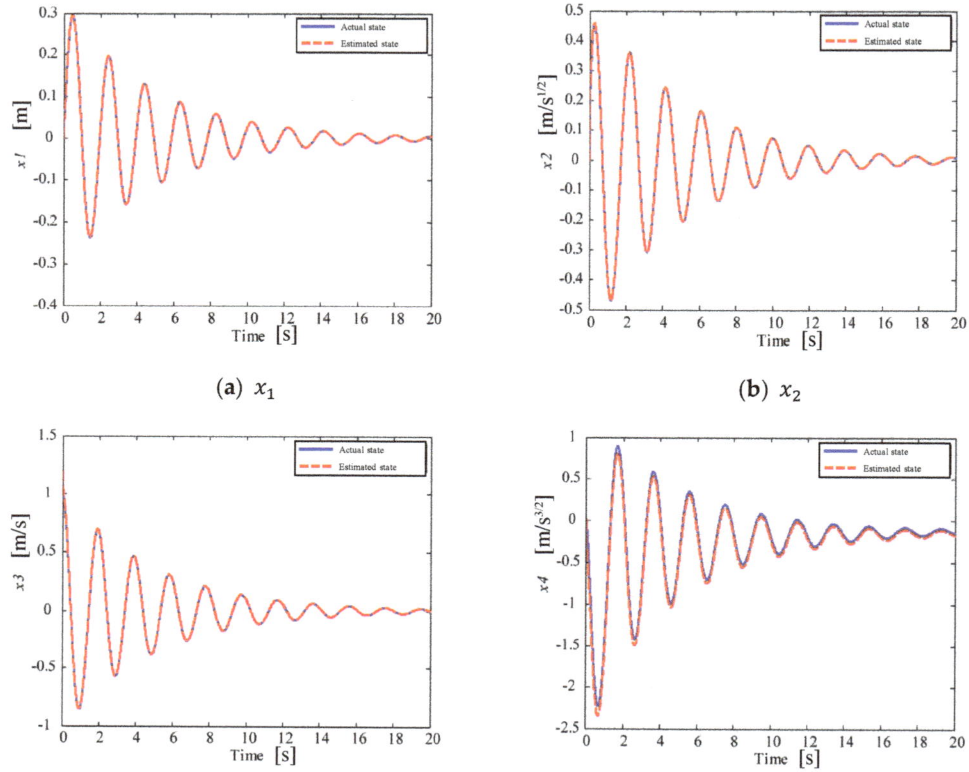

(**a**) x_1 (**b**) x_2 (**c**) x_3 (**d**) x_4

Figure 6. Results for actual (blue lines) and estimated (red broken lines) states.

Next, fractional-order LQR control is conducted. In comparing the states with control and those without control, the gaps between the actual states and the estimated values are shown in Figure 7, and the state evolution is illustrated in Figure 8. The results with control are obtained with output feedback control using a fractional-order LQR.

Figure 7 confirms that the estimation of the states by the fractional-order state observer is performed successively. Judging from the results in Figure 8, compared with the no-control case, good damping effects are confirmed in the case that fractional-order LQR control is applied.

Consequently, using the feedback gains obtained with the iteration-based method and the fractional-order state observer, it is proven that fractional-order LQR control can be carried out for a controlled object that includes fractional-order derivative states in its state equation.

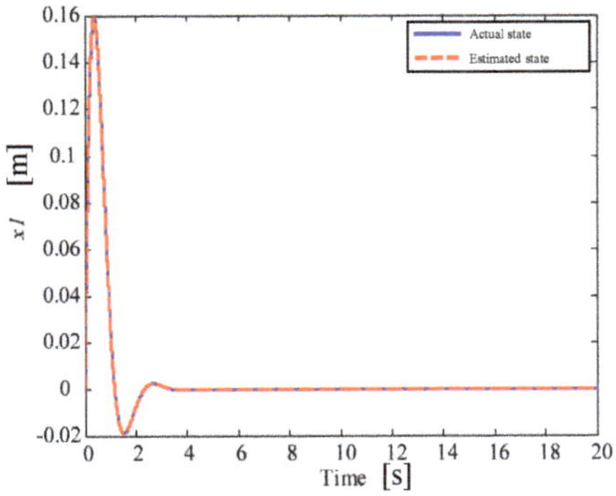

(a) x_1

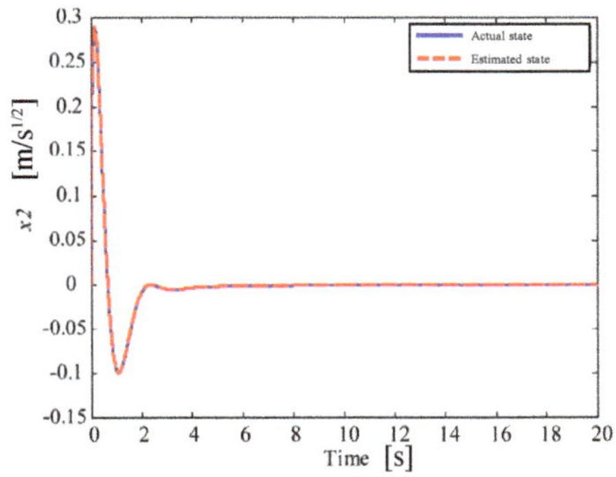

(b) x_2

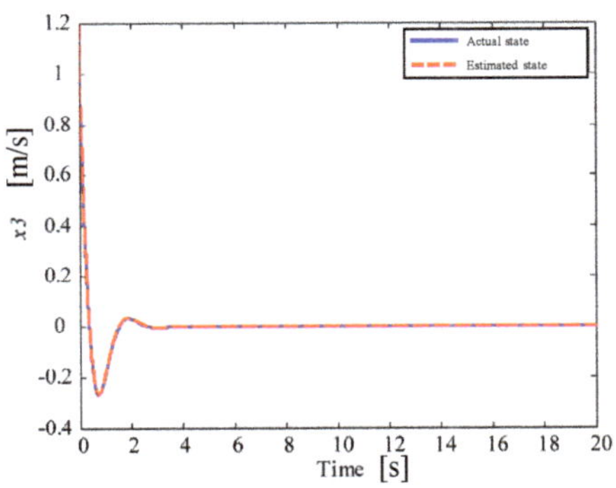

(c) x_3

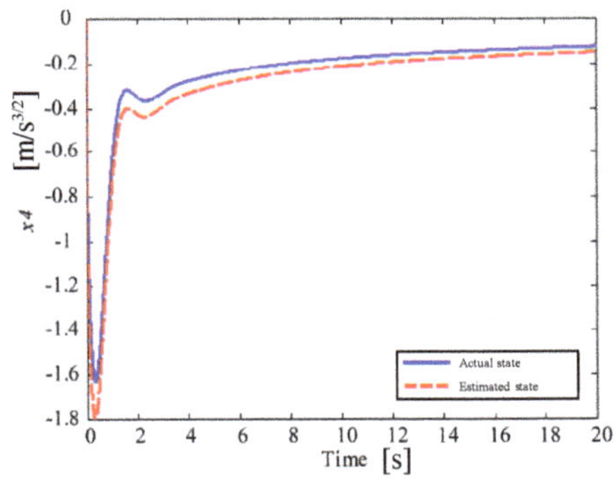

(d) x_4

Figure 7. Difference between actual (blue lines) and estimated (red broken lines) states.

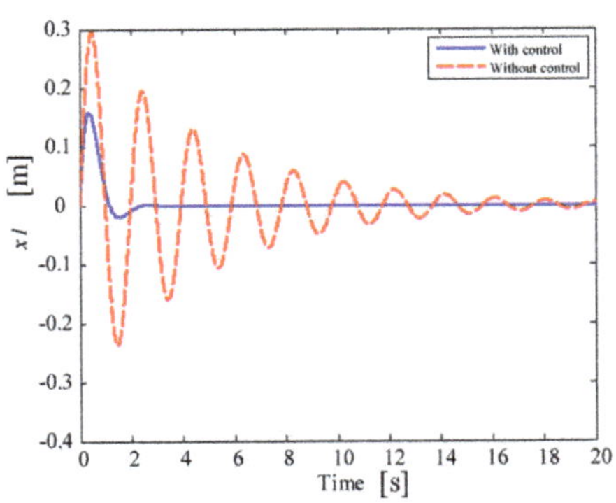

(a) x_1

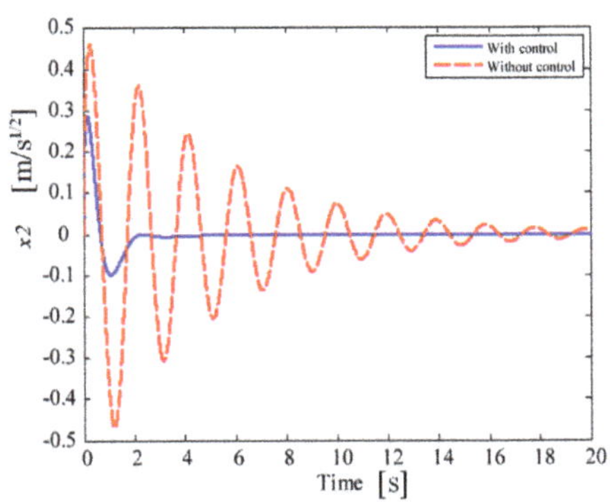

(b) x_2

Figure 8. *Cont.*

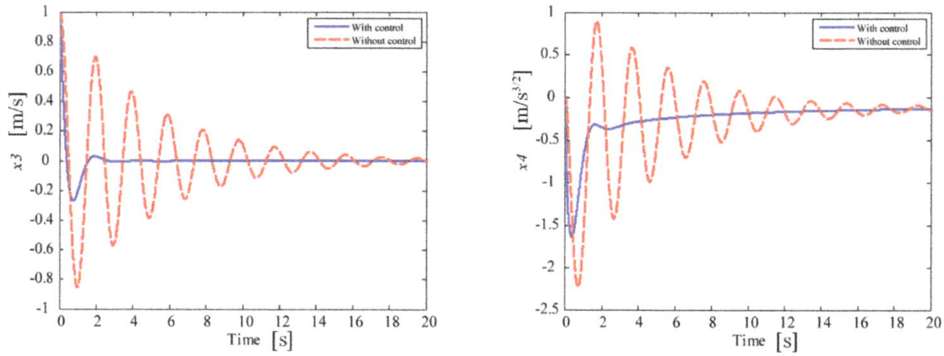

(c) x_3 (d) x_4

Figure 8. Results for states with/without fractional-order LQR control (with control: blue lines, without control: red broken lines).

7.2. Comparison between Proposed Method and Sierociuk and Vinagre Method

Moreover, the control results for the fractional-order LQR using the feedback gains obtained using the proposed method and the fractional-order LQR using the feedback gains obtained with the Sierociuk and Vinagre method [19] are compared. Using the same weighting matrices, $\boldsymbol{Q}$ and $\boldsymbol{R}$ in Equation (60), the feedback gains are calculated with the Sierociuk and Vinagre method as follows:

$$\boldsymbol{F}_{SV} = \begin{bmatrix} 0.5394 & 14.7824 & 13.0640 & 5.1116 \end{bmatrix}. \tag{69}$$

The poles of the closed-loop system-matrix $\boldsymbol{A} - \boldsymbol{B}\boldsymbol{F}_{SV}$ are as follows:

$$Poles = \begin{cases} -1.5649 \pm 1.5733i \\ -0.9908 \pm 0.9774i \end{cases}. \tag{70}$$

These poles are located in region ③ in Figure 1. As a result, the controlled states with the feedback gains in Equation (61) and the controlled states with the feedback gains in Equation (69) can be compared, as shown in Figure 9. The results with control are obtained with output feedback control using a fractional-order LQR.

From Figure 9, it is clear that both feedback gains exhibit good control effects. In particular, in the vibration waveforms controlled using the feedback gains obtained with the Sierociuk and Vinagre method, all of the states are confirmed to be overdamped. However, it is obvious that the convergence of x_1 (displacement) to the target value is considerably slow.

Next, the design of the weighting matrices $\boldsymbol{Q}$ and $\boldsymbol{R}$ in the quadratic cost function J is modified as follows:

$$\boldsymbol{Q} = \begin{bmatrix} 10 & 0 & 0 & 0 \\ 0 & 10 & 0 & 0 \\ 0 & 0 & 10 & 0 \\ 0 & 0 & 0 & 10 \end{bmatrix}, \boldsymbol{R} = 1. \tag{71}$$

In this case, the weights in $\boldsymbol{Q}$ are placed on not only the integer-order derivative states, but also the fractional-order derivative states. Using weighting matrices $\boldsymbol{Q}$ and $\boldsymbol{R}$, the feedback gain matrix derived using the proposed method is as follows:

$$\boldsymbol{F} = \begin{bmatrix} 0.5394 & -2.3077 & 3.7258 & -0.9831 \end{bmatrix}. \tag{72}$$

The poles of the closed-loop system matrix $\boldsymbol{A}-\boldsymbol{BF}$ are as follows:

$$Poles = \left\{ \begin{array}{l} -1.0806 \pm 1.5848i \\ -0.5891 \pm 1.4985i \end{array} \right. . \tag{73}$$

These poles are located in region ② in Figure 1. On the other hand, the feedback gain matrix obtained by the Sierociuk and Vinagre method is as follows:

$$\boldsymbol{F}_{SV} = \left[\begin{array}{cccc} 0.5394 & 16.2648 & 15.1137 & 6.3425 \end{array} \right]. \tag{74}$$

The poles of the closed-loop system matrix $\boldsymbol{A}-\boldsymbol{BF}_{SV}$ are as follows:

$$Poles = \left\{ \begin{array}{l} -2.3380 \pm 0.3989i \\ -0.8332 \pm 1.0007i \end{array} \right. . \tag{75}$$

These poles are located in region ③ in Figure 1. The states controlled with the above-mentioned feedback-gain matrices are compared in Figure 10. The results with control are obtained with output feedback control using a fractional-order LQR.

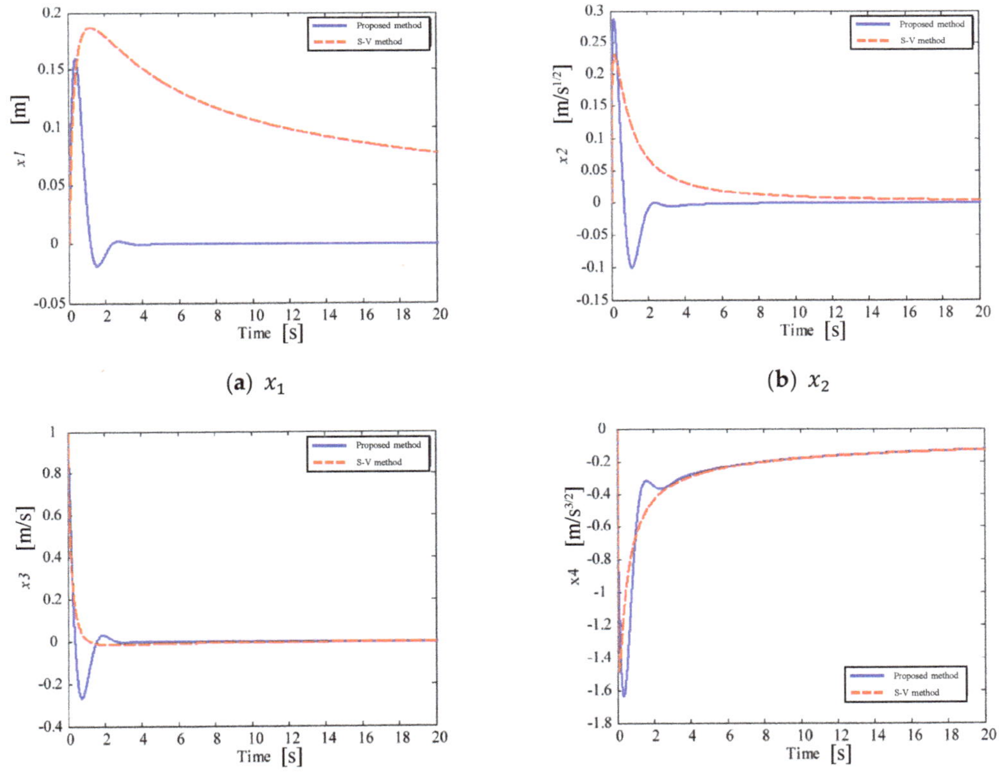

(**a**) x_1 (**b**) x_2 (**c**) x_3 (**d**) x_4

Figure 9. Results for states with fractional-order LQR control (proposed method: blue lines, Sierociuk and Vinagre method (S–V method): red broken lines).

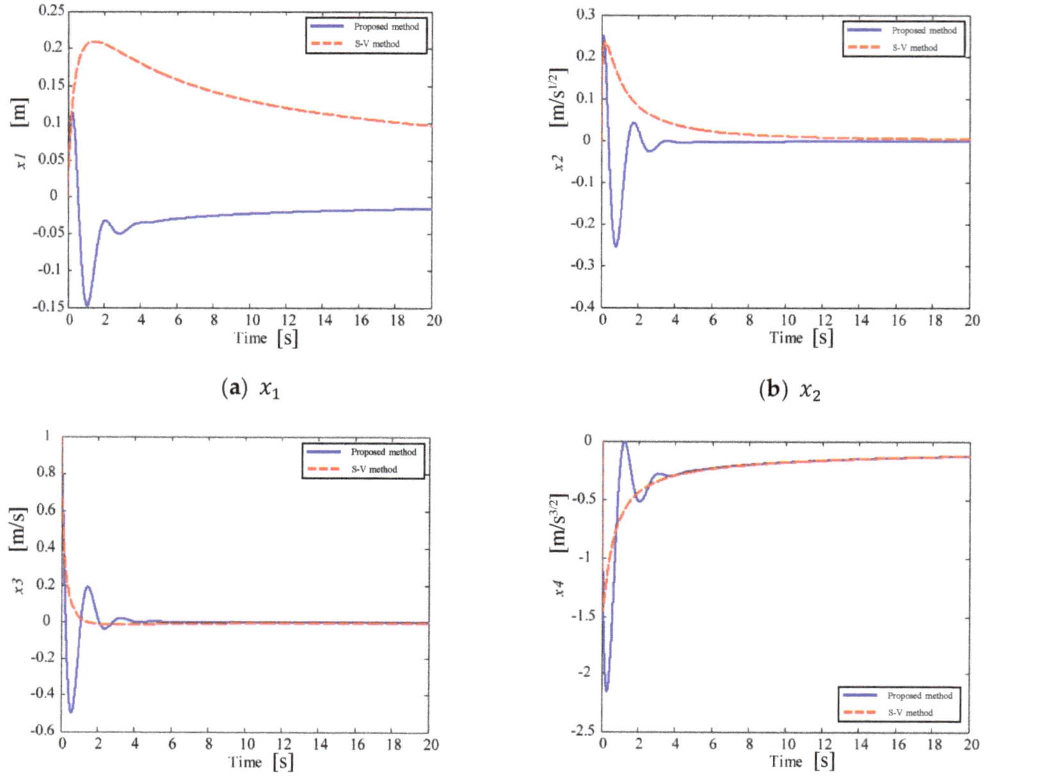

(a) x_1 (b) x_2 (c) x_3 (d) x_4

Figure 10. Results for states with fractional-order LQR control (proposed method: blue lines, Sierociuk and Vinagre method (S–V method): red broken lines).

Figure 10 demonstrates that the results of the states are oscillations damped by the feedback-gain matrix $\boldsymbol{F}$ and are overdamped by the feedback-gain matrix $\boldsymbol{F}_{SV}$. Furthermore, the convergence property of x_1 in Figure 10 is similar to that in Figure 9. However, in the proposed method, the vibration suppression effect becomes worse when the weights in $\boldsymbol{Q}$ are placed on the fractional-order derivative states. On the other hand, in the Sierociuk and Vinagre method, the weights on the fractional-order derivative states in $\boldsymbol{Q}$ can play an important role for better vibration control effects.

8. Conclusions

The present study investigated whether the LQR control method could be applied to a state equation with fractional-order derivative states. A design method for a fractional-order LQR control system was proposed in order to deal with a controlled object that included fractional-order derivative terms in the equation of motion.

First, in order to explain the design method, a viscoelastic damper was modeled by an equation of motion including a fractional-derivative term. The equation of motion was able to be transformed into a fractional-order state equation.

Second, a fractional-order LQR was applied to control the dynamics of the viscoelastically damped structure. A design method for fractional-order LQR control was proposed using the Lyapunov equation and Lagrange's method for undetermined multipliers. A new iteration-based approach to solve the algebraic Riccati equation was established in order to obtain the feedback gains for the fractional-order LQR.

Third, a fractional-order state observer was constructed in order to estimate the fractional-order derivative states. All states for the viscoelastic damper system were successfully estimated using the fractional-order state observer.

Fourth, a numerical calculation algorithm corresponding to a differential equation containing fractional derivative terms was introduced. The numerical calculation algorithm was based on the Grünwald–Letnikov definition of fractional differentiation. Numerical calculation results using the introduced method and the exact solution of the viscoelastic damper model were compared in order to investigate its calculation accuracy.

Finally, using the introduced numerical simulation method, LQR control effect simulations were carried out. The numerical simulations revealed that the iterative method was valid for obtaining the feedback gains for the fractional-order LQR. The fractional-order LQR controller was confirmed to be capable of successfully suppressing vibrations appearing in a vibratory system with viscoelasticity. Moreover, the fractional-order LQR obtained by the Sierociuk and Vinagre method and that derived by the proposed method were compared with respect to their control effects.

However, the fractional-order state observer in the present paper is not optimal, and therefore needs to be improved. In addition, control effects by a fractional-order LQR to suppress vibrations appearing in a vibratory system with viscoelasticity must be compared with control effects by an integer-order LQR.

Author Contributions: Conceptualization, M.K.; methodology, A.T., T.Y. and N.K.; software, A.T. and N.K.; validation, A.T.; formal analysis, A.T. and T.Y.; investigation, A.T. and M.K.; resources, N.K. and M.K.; data curation, A.T. and T.Y.; writing—original draft preparation, A.T. and M.K.; writing—review and editing, M.K.; visualization, A.T.; supervision, N.K. and M.K.; project administration, M.K. All authors have read and agreed to the published version of the manuscript.

Funding: This research received no external funding.

Institutional Review Board Statement: Not applicable.

Informed Consent Statement: Not applicable.

Data Availability Statement: Not applicable.

Conflicts of Interest: The authors declare no conflict of interest.

References

1. Oldham, K.B.; Spanier, J. *The Fractional Calculus*; Dover: Mineola, NY, USA, 2006.
2. Podlubny, I. Fractional-order systems and $PI^{\lambda}D^{\mu}$-controllers. *IEEE Trans. Autom. Control* **1999**, *44*, 208–214. [CrossRef]
3. Cajo, R.; Muresan, C.I.; Ionescu, C.M.; De Keyser, R.; Plaza, D. Multivariable Fractional Order PI Autotuning Method for Heterogeneous Dynamic Systems. In Proceedings of the 3rd IFAC Conference on Advances in Proportional-Integral-Derivative Control, Ghent, Belgium, 9–11 May 2018; pp. 865–870.
4. Muresan, C.I.; Copot, C.; Birs, I.; De Keyser, R.; Vanlanduit, S.; Ionescu, C.M. Experimental Validation of a Novel Auto-Tuning Method for a Fractional Order PI Controller on an UR10 Robot. *Algorithms* **2018**, *11*, 95. [CrossRef]
5. Folea, S.; De Keyser, R.; Birs, I.R.; Muresan, C.I.; Ionescu, C. Discrete-Time Implementation and Experimental Validation of a Fractional Order PD Controller for Vibration Suppression in Airplane Wings. *Acta Polytech. Hung.* **2017**, *14*, 191–206.
6. Birs, I.R.; Folea, S.; Copot, D.; Prodan, O.; Muresan, C.I. Comparative analysis and experimental results of advanced control strategies for vibration suppression in aircraft wings. *J. Phys. Conf. Ser.* **2017**, *783*, 012054. [CrossRef]
7. Li, H.; Luo, Y.; Chen, Y. A Fractional Order Proportional and Derivative (FOPD) Motion Controller: Tuning Rule and Experiments. *IEEE Trans. Control Syst. Technol.* **2010**, *18*, 516–520. [CrossRef]
8. Swain, S.K.; Sain, D.; Mishra, S.K.; Ghosh, S. Real time implementation of fractional order PID controllers for a magnetic levitation plant. *AEU Int. J. Electron. Commun.* **2017**, *78*, 141–156. [CrossRef]
9. Birs, I.R.; Muresan, C.I.; Prodan, O.; Folea, S.C.; Ionescu, C. Structural vibration attenuation using a fractional order PD controller designed for a fractional order process. *IFAC-PapersOnLine* **2018**, *51*, 533–538. [CrossRef]
10. Muresan, C.I. Simplified Optimization Routine for Tuning Robust Fractional Order Controllers. *Am. J. Comput. Math.* **2013**, *3*, 7–12. [CrossRef]

11. Birs, I.R.; Muresan, C.I.; Folea, S.; Prodan, O.; Kovacs, L. Vibration suppression with fractional-order $PI^{\lambda}D^{\mu}$ controller. In Proceedings of the 2016 IEEE International Conference on Automation, Quality and Testing, Robotics (AQTR), Cluj-Napoca, Romania, 19–21 May 2016; pp. 1–6.
12. Tejado, I.; Vinagre, B.M.; Traver, J.E.; Prieto-Arranz, J.; Nuevo-Gallardo, C. Back to Basics: Meaning of the Parameters of Fractional Order PID Controllers. *Mathematics* **2019**, *7*, 530. [CrossRef]
13. Li, C.; Zhang, N.; Lai, X.; Zhou, J.; Xu, Y. Design of a fractional-order PID controller for a pumped storage unit using a gravitational search algorithm based on the Cauchy and Gaussian mutation. *Inf. Sci.* **2017**, *396*, 162–181. [CrossRef]
14. Zamani, A.; Barakati, S.M.; Yousofi-Darmian, S. Design of a fractional order PID controller using GBMO algorithm for load–frequency control with governor saturation consideration. *ISA Trans.* **2016**, *64*, 56–66. [CrossRef] [PubMed]
15. Padula, F.; Visioli, A. Tuning rules for optimal PID and fractional-order PID controllers. *J. Process Control* **2011**, *21*, 69–81. [CrossRef]
16. Dumlu, A.; Erenturk, K. Trajectory Tracking Control for a 3-DOF Parallel Manipulator Using Fractional-Order PIlDm Control. *IEEE Trans. Ind. Electron.* **2014**, *61*, 3417–3426. [CrossRef]
17. Zeng, G.-Q.; Chen, J.; Dai, Y.-X.; Li, L.-M.; Zheng, C.-W.; Chen, M.-R. Design of fractional order PID controller for automatic regulator voltage system based on multi-objective extremal optimization. *Neurocomputing* **2015**, *160*, 173–184. [CrossRef]
18. Chen, Z.; Yuan, X.; Ji, B.; Wang, P.; Tian, H. Design of a fractional order PID controller for hydraulic turbine regulating system using chaotic non-dominated sorting genetic algorithm II. *Energy Convers. Manag.* **2014**, *84*, 390–404. [CrossRef]
19. Sierociuk, D.; Vinagre, B.M. Infinite Horizon State-feedback LQR Controller for Fractional Systems. In Proceedings of the 49th IEEE Conference on Decision and Control, Atlanta, GA, USA, 15–17 December 2010; pp. 6674–6679.
20. Das, S.; Pan, I.; Halder, K.; Das, S.; Gupta, A. LQR based improved discrete PID controller design via optimum selection of weighting matrices using fractional order integral performance index. *Appl. Math. Model.* **2013**, *37*, 4253–4268. [CrossRef]
21. Li, Y.; Chen, Y.Q. Fractional Order Linear Quadratic Regulator. In Proceedings of the 2008 IEEE/ASME International Conference on Mechatronic and Embedded Systems and Applications, Beijing, China, 12–15 October 2008; pp. 363–368.
22. Birs, I.; Folea, S.; Ionescu, F.; Prodan, O.; Muresan, C. Preliminary results and simulation of an active pendulum system for a three floor building. *Procedia Eng.* **2017**, *199*, 1647–1652. [CrossRef]
23. Sumathi, R.; Umasankar, P. Optimal design of fractional order PID controller for time-delay systems: An IWLQR technique. *Int. J. Gen. Syst.* **2018**, *47*, 714–730. [CrossRef]
24. Hartley, T.T.; Lorenzo, C.F. Dynamics and Control of Initialized Fractional-Order Systems. *Nonlinear Dyn.* **2002**, *29*, 201–233. [CrossRef]
25. Tejado, I.; Vinagre, B.M.; Sierociuk, D. State space methods for fractional controllers design. In *Handbook of Fractional Calculus with Applications, Vol. 6: Applications in Control*; Petráš, I., Ed.; De Gruyter: Berlin, Germany, 2019; pp. 175–199.
26. Monje, C.A.; Chen, Y.Q.; Vinagre, B.M.; Xue, D.; Feliu, V. *Fractional-Order Systems and Controls*; Springer: London, UK, 2010.
27. Bagley, R.L.; Torvik, P.J. A Theoretical Basis for the Application of Fractional Calculus to Viscoelasticity. *J. Rheol.* **1983**, *27*, 201–210. [CrossRef]
28. Shimizu, N.; Iijima, M. Fractional Differential Model of Viscoelastic Material. *Trans. JSME C* **1996**, *62*, 4447–4451. (In Japanese) [CrossRef]
29. Ikeda, F.; Kawata, S.; Watanabe, A. An Optimal Regulator Design of Fractional Differential Systems. *Trans. Soc. Instrum. Control Eng.* **2001**, *37*, 856–861. (In Japanese) [CrossRef]
30. Kodama, S.; Suda, N. *Matrix Theory for Systems and Control*; SICE, Ed.; Corona: Tokyo, Japan, 1981. (In Japanese)
31. Dadras, S.; Momeni, H.R. A New Fractional Order Observer Design for Fractional Order Nonlinear Systems. In Proceedings of the ASME 2011 International Design Engineering Technical Conferences and Computers and Information in Engineering Conference, Washington, DC, USA, 28–31 August 2011; Volume 3, pp. 403–408.
32. Matignon, D.; d'Andréa-Novel, B. Observer-based controllers for fractional differential systems. In Proceedings of the 36th Conference on Decision and Control, San Diego, CA, USA, 10–12 December 1997; pp. 4967–4972.
33. Zhang, W.; Shimizu, N. Numerical Algorithm for Dynamic Problems Involving Fractional Operators. *JSME Int. J. Ser. C* **1998**, *41*, 364–370.
34. Fukunaga, M.; Shimizu, N. A high-speed algorithm for computation of fractional differentiation and fractional integration. *Phil. Trans. R. Soc. A* **2013**, *371*, 20120152. [CrossRef]
35. Chen, Y.Q.; Petráš, I.; Xue, D. Fractional Order Control—A Tutorial. In Proceedings of the 2009 American Control Conference, St. Louis, MO, USA, 10–12 June 2009; pp. 1397–1411.
36. Xue, D. *Fractional-Order Control Systems*; De Gruyter: Berlin, Germany, 2017.
37. Ikeda, F.; Kawata, S.; Oguchi, T. Vibration Control of Flexible Structures with Fractional Differential Active Mass Dampers. *Trans. JSME C* **2001**, *67*, 2798–2805. (In Japanese) [CrossRef]

MDPI
St. Alban-Anlage 66
4052 Basel
Switzerland
Tel. +41 61 683 77 34
Fax +41 61 302 89 18
www.mdpi.com

Applied Sciences Editorial Office
E-mail: applsci@mdpi.com
www.mdpi.com/journal/applsci